SKETCHBOOK TO ACCOMPANY

Engineering Drawing

A.W. BOUNDY

AssDipMechEng, MPhil

Mc
Graw
Hill

Boston   Burr Ridge, IL   Dubuque, IA   Madison, WI   New York
San Francisco   St. Louis   Bangkok   Bogotá   Caracas   Kuala Lumpur
Lisbon   London   Madrid   Mexico City   Milan   Montreal   New Delhi
Santiago   Seoul   Singapore   Sydney   Taipei   Toronto

McGraw·Hill Australia

A Division of The **McGraw·Hill** Companies

Reprinted 2002, 2003, 2004 (twice), 2005, 2006
Copyright © 2002 McGraw-Hill Australia Pty Limited
Additional owners of copyright are named in on-page credits.

Boundy, A. W (Albert William)

Sketchbook to accompany Engineering Drawing 6e

Part no. 777777693–2 (Book/Sketchbook set ISBN 0 07 471043 5)

Published in Australia by
McGraw-Hill Australia Pty Limited
Level 2, 82 Waterloo Road, North Ryde, NSW 2113, Australia
Sponsoring Editor: Michael Tully
Production Manager: Jo Munnelly
Marketing Manager: Sara Dougall
Editor: Catherine Dunk
Designer: Ramsay Macfarlane
Typesetters: Post Pre-Press Group
Proofreader: Trish Fox
Illustrator: Alan Laver, Shelley Communications
Printed by: Kyodo Printing Co. (Singapore) Pte Ltd

The art of freehand sketching is one that all tradespersons, draftspersons, technicians and professional engineers will have to develop to some degree of competence very early in their careers.

Communication in engineering often involves transferring ideas of shape, size, form and dimension to clients and others within the profession. Engineering drawings are normally the final outcome of the design process and form part of the specification and contract documents between the client and engineer. However, in the lead up to the production of engineering drawings, engineers, draftspersons and tradespersons generally produce numerous freehand sketches to assess formative ideas before reaching a final decision. The ability to produce a good freehand sketch is, therefore, a desirable asset when discussing final configurations of designs leading up to the production of detailed drawings.

Types of freehand sketches

Types of sketches can vary from so-called rough 'mud maps' to accurate detailed sketches produced on prepared sketching paper similar to that used in this book. However, the type selected should convey to the recipient of the sketch sufficient information to enable this person to proceed into the next stage of the design process.

The drawer of the sketch must decide if it should be two- or three-dimensional and whether it should be dimensioned or not. The engineering experience of the drawer will determine the choice of what is done.

Two-dimensional sketches are normally multiview sketches based on the principles of orthogonal projection. This type of sketch is the simplest to draw and can easily include dimensions, but it lacks the 'true to life' look provided by a good three-dimensional sketch. The latter, however, is harder to draw and, unless it is a reasonable attempt, can fail to adequately provide the 'true to life' appearance.

Sketching materials

An ideal medium on which to produce a freehand sketch is prepared sketching paper, either orthogonal or isometric depending on the sketch to be prepared. If available, it should always be used, and a supply should be provided if it is known that sketches will have to be produced on the site or at a meeting. Sometimes it is necessary to use normal ruled paper, but this makes the task of producing a good sketch much harder.

Other essential sketching materials are an HB (or darker) 0.5 mm clutch pencil and a white eraser. A ruler, while not essential, is helpful when setting out views. The pencil must be soft and capable of producing a dark line easily. It is frustrating trying to read a sketch that is faintly drawn and does not readily distinguish between outlines, hidden detail and dimension lines.

Two-dimensional sketches

Two-dimensional sketches normally comprise one or more orthogonal views of a component and may or may not be accompanied by dimensions. It is just as important when sketching orthogonal views as when drawing them to place them relative to each other according to the principles of third-angle projection explained in textbooks.

Orthogonal views (front, top, right and left side) generally comprise a variety of straight lines, circles and arcs. Centre lines (long dash, short dash), hidden detail lines (short dashes) and dimension lines (light lines) are necessary to complete these views. When using prepared sketching paper the length of straight lines is easily assessed if the dots on the sketching paper are taken as either 5 mm or 7 mm apart, depending on the direction being measured. The scale of distance is shown in the top right-hand corner of each page.

Sketching techniques

Individuals usually will develop their own techniques when sketching straight lines and circles: the finished sketch is the important outcome of the exercise, not how it is done. However, a few hints will help to point beginners in the right direction for developing their own skills.

The easiest line to draw is a left to right line drawn horizontally across the page. Fortunately, all straight lines can be made left to right and horizontal simply by turning the sketching book around until it is in the correct position. If there is adequate time it is a good idea to *lightly* sketch the whole drawing and then, when you are sure everything is correct, proceed with final lining in using a single heavy stroke of the pencil. The reason for drawing a light outline first is that you will make frequent errors and it only requires a light touch of the rubber to erase light lines.

The hardest line to sketch is the circle or arc and you will have to develop the skill to produce shapes that look reasonably like circles. Following these simple steps will help you to produce a circle that is satisfactory. (It is the method the author finds best for drawing the circle outline, but it may not necessarily suit everyone.)

1. Find the centre of the circle and if it is a small circle (Ø20 mm or less) find four points on the circumference 90 degrees apart (Fig.1(a)).
2. Commence at the left diametral point and *very slowly* draw a light circular arc clockwise to the top diametral point. Then from the right diametral point very slowly draw a light circular arc clockwise to the bottom diametral point. Turn the paper through 90 degrees and repeat the two arcs to complete the circle. If the circle looks true, line it in slowly. If it is out of shape, erase the offending line and try again. Left handers should start at the right diametral point and move anti-clockwise.
3. For circles greater than Ø20 mm select the four diametral points as in point l, plus four more

diagonal (45 degrees) diametral points (Fig.1(b)). The distance for the diagonal diametral points can be easily assessed as the diagonal distance between points is 7 mm.
4. Proceed as in point 2, ensuring that the circular arcs pass through the diagonal points in a smooth curve.

Sketching isometric and oblique circles

Circles and part circles requiring construction in isometric sketches are all contained within the framework of a rhombus (that is, a square pushed out of shape). However, in oblique sketching the front face is not distorted and circular detail contained within this view remain as circles, while circles contained on the top or either end face are distorted and contained within the framework of a rhomboid (that is, a rectangle pushed out of shape).

The result of the distortion in both types of pictorial sketching is that circles become ellipses and need to be plotted if some degree of accuracy and correct shape is to be achieved. Figure 2(a) on p. vi illustrates circles and part circles found on isometric sketches. There are three types of isometric circles (ellipses) and these are illustrated on the sketch

FIGURE I

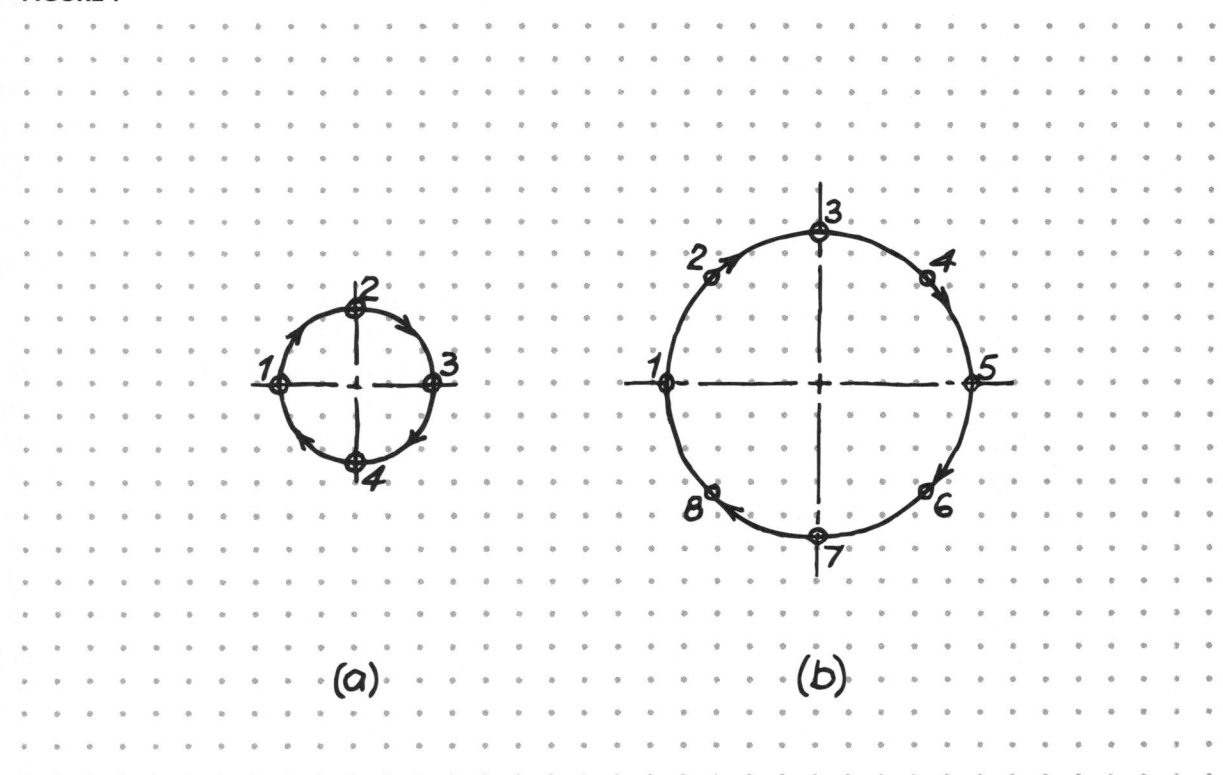

(a)   (b)

marked 'circle types'. The circles are identical but have different orientations according to which isometric face they lie on. A small circle can easily be outlined within its rhombus by drawing the dividers of each side to intersect at the centre of the rhombus. Where the dividers intersect the side gives four points on the ellipse. For a large circle eight points are required and after drawing in the side dividers to give four points E, F, G and H, the two diagonals AC and BD are drawn and they too intersect at the centre O of the rhombus. Join EH and FG and divide that part of the long diagonal equally, from A and C on to EH and FG, respectively, to give two more points on the ellipse. The other two points are found by dividing BO and DO in the ratio of 1:2.

Corners are plotted using two points found by drawing the particular corner rhombus (Fig.2(a)). A rounded end requires two corner rhombuses and five points as shown.

Oblique circles are much narrower than isometric circles, being plotted in rhomboids having their short side equal to half the length of the long side (Fig.2(b)). Similarly to isometric circles, the small oblique circle is sketched using four points, found by drawing the side dividers. Sketching the oblique ellipse is somewhat harder than sketching the isometric ellipse. It is narrower and its long diagonal XX does not coincide with HF or lie on the long diagonal BD of the rhomboid, but is midway between these two and is not normally plotted. Similarly, the short diagonal YY is at right angles to YY. Once this is realized, drawing the curve becomes much easier. A large oblique circle requires eight points to draw it smoothly. Four points are obtained by drawing the side dividers EG and HF and then the long and short diagonals of the rhomboid, BD and AC, to intersect at O. The next four points are found by dividing HO and FO in the ratio of 2:5 as shown. Through these points, draw lines parallel to AD and BC to intersect the long and short diagonals to give four more points. Remember the secret of sketching this ellipse is to visualize the positions of XX and YY about which the ellipse is symmetrical.

Of course many sketchers avoid views which entail having circular or cylindrical detail on the oblique faces, preferring to orientate the drawing so that this type of detail is on the front face where it is circular. Sometimes, however, this cannot be avoided, when circular or cylindrical detail is found on both the front and side faces as with the exercise on page 30.

Corners and rounded ends are drawn using two and five points, respectively, and the appropriate construction rhomboid as shown in Fig. 2(b).

Construction lines used in sketching should be lightly drawn and thus easily removed with a sharp edged rubber. Lining in should be a slow and deliberate process with the sketch pad being turned frequently to follow the line or curve required in the most comfortable manner.

Making best use of the book

This book has been designed to serve a two-fold purpose:

1. It provides a complete course in sketching orthogonal and pictorial views.
2. It tests and improves your ability to interpret drawings by converting from orthogonal to pictorial views and vice versa. This is especially important if you have not been introduced to these topics during your secondary school studies.

Answers are supplied for all the exercises in the book. The exercises are set on one page and answered on the next page (or the previous page where appropriate). You should attempt the exercises before turning the page to check the answers. At the back of the book are several pages with grids only. They may be photocopied so that you can attempt exercises of your own or your instructor's choice.

Complete all the exercises in this course and you should have achieved a reasonable level of knowledge in drawing interpretation as well as a satisfactory level of sketching skills. Good sketching, and rest assured that your level of competence will be commensurate with the effort you put into the exercises.

Note: Paper size in the illustrations has been reduced to slightly smaller than true size. You should use the scales as indicated.

FIGURE 2

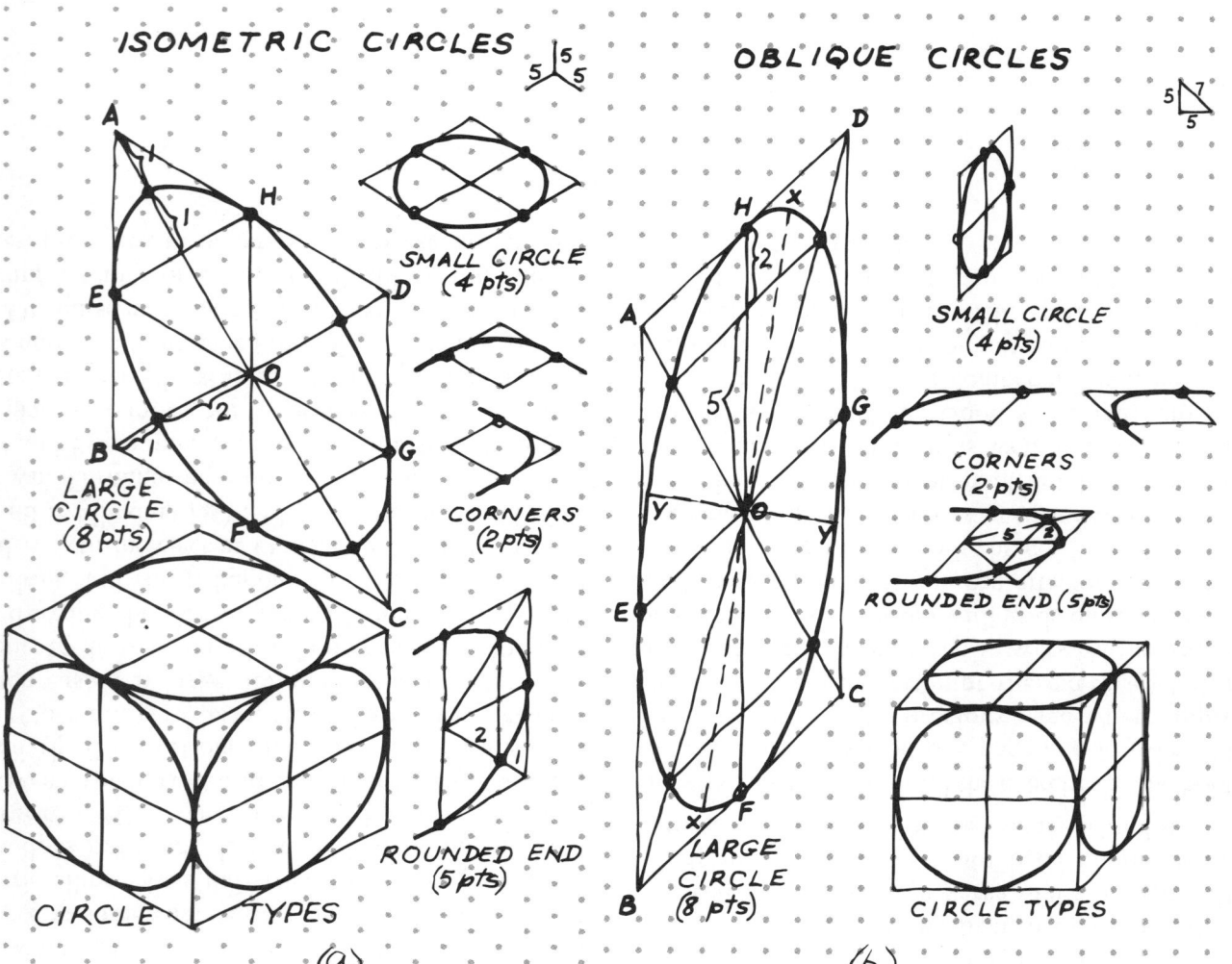

ISOMETRIC CIRCLES

LARGE CIRCLE (8 pts)

SMALL CIRCLE (4 pts)

CORNERS (2 pts)

ROUNDED END (5 pts)

CIRCLE TYPES

OBLIQUE CIRCLES

LARGE CIRCLE (8 pts)

SMALL CIRCLE (4 pts)

CORNERS (2 pts)

ROUNDED END (5 pts)

CIRCLE TYPES

(a)

(b)

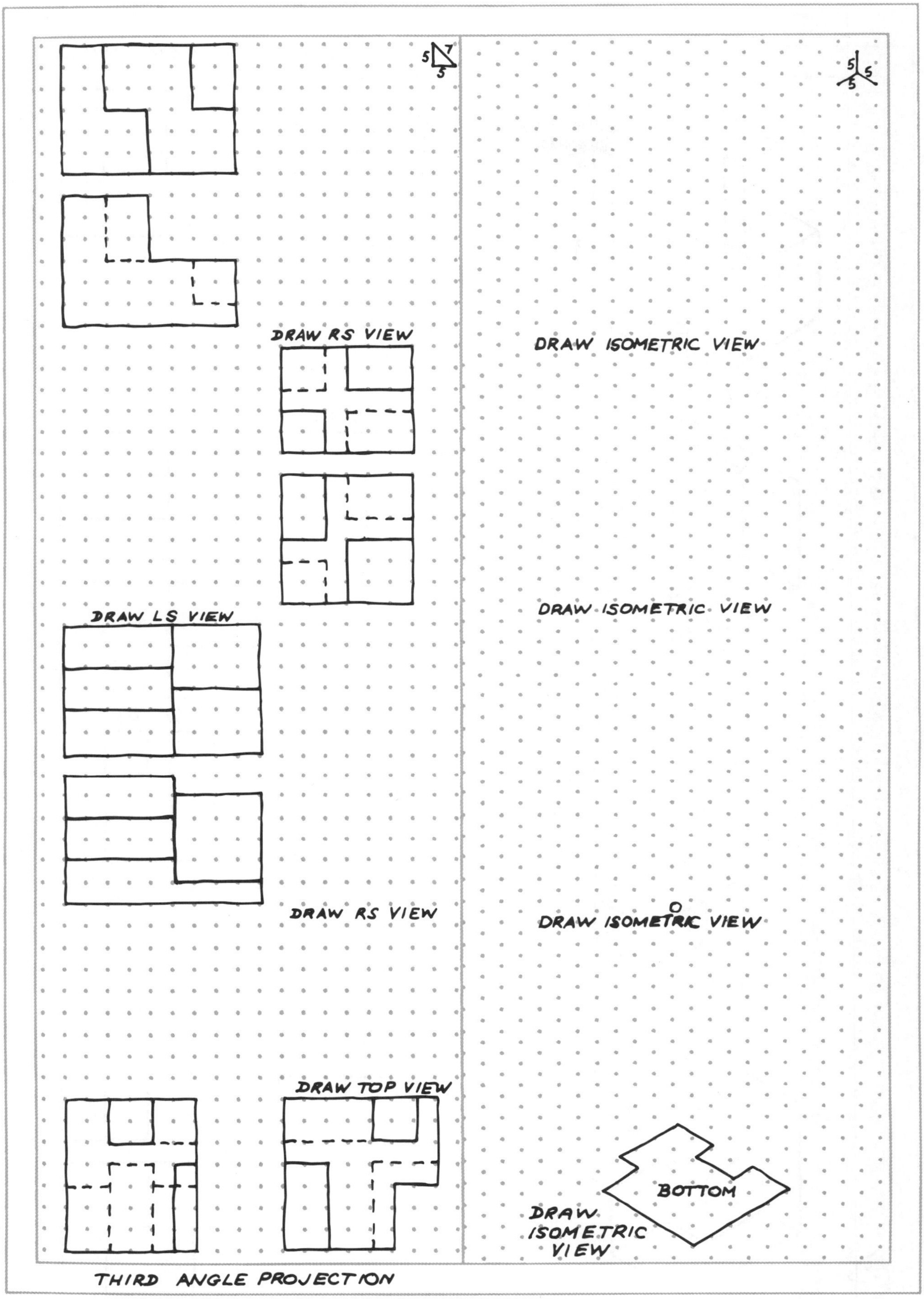

DRAW RS VIEW

DRAW ISOMETRIC VIEW

DRAW ISOMETRIC VIEW

DRAW LS VIEW

DRAW RS VIEW

DRAW ISOMETRIC VIEW

DRAW TOP VIEW

BOTTOM

DRAW
ISOMETRIC
VIEW

THIRD ANGLE PROJECTION

1

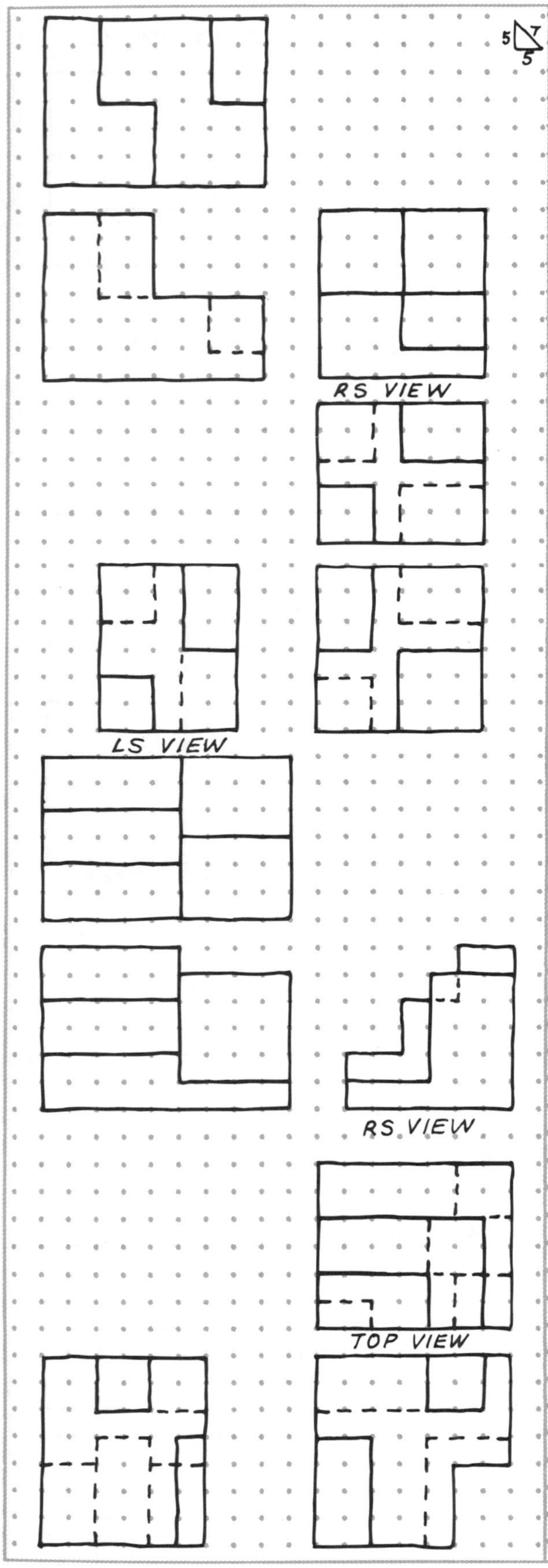

5 7
5

5 5
5

RS VIEW

LS VIEW

RS VIEW

TOP VIEW

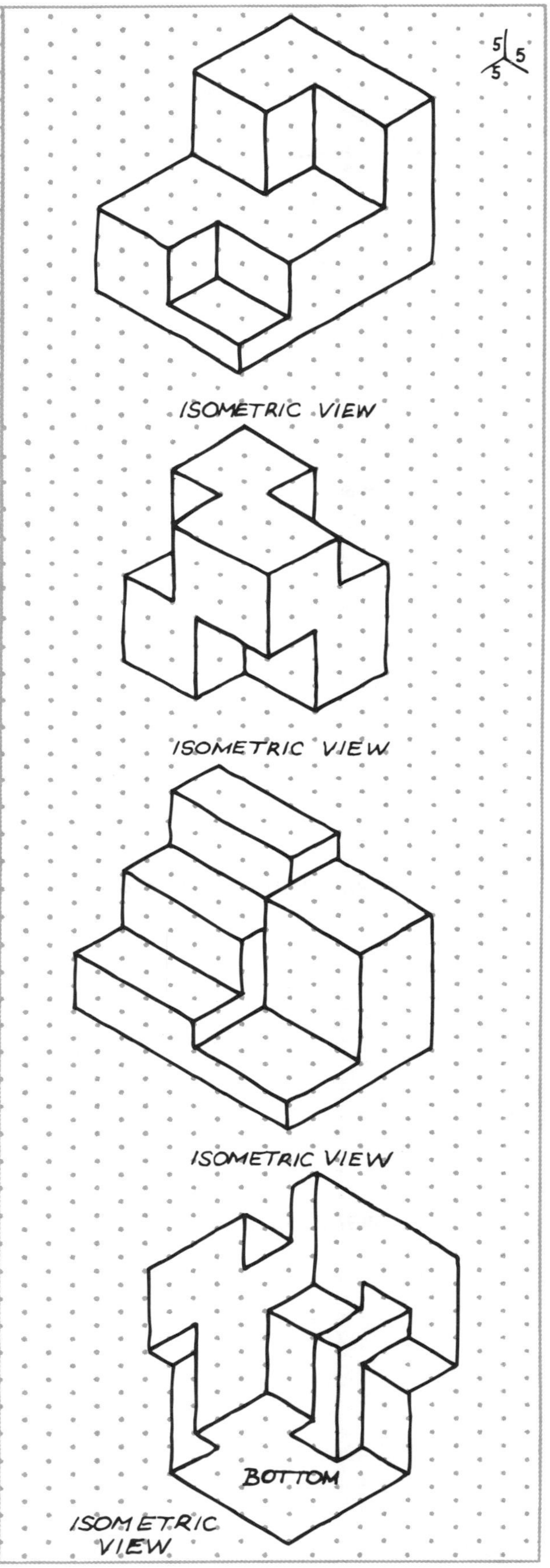

ISOMETRIC VIEW

ISOMETRIC VIEW

ISOMETRIC VIEW

BOTTOM

ISOMETRIC VIEW

THIRD ANGLE PROJECTION

2

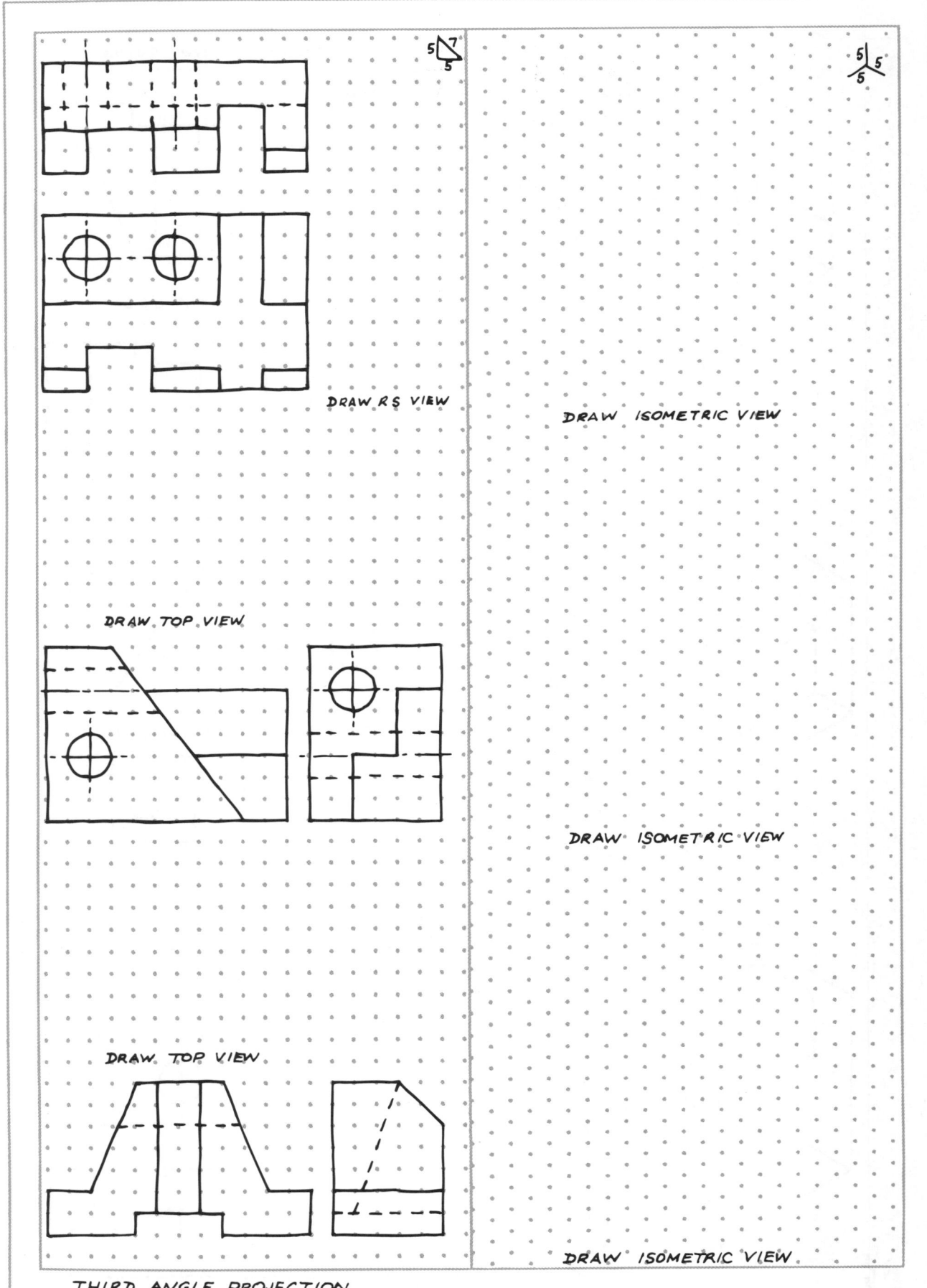

DRAW R S VIEW

DRAW ISOMETRIC VIEW

DRAW TOP VIEW

DRAW ISOMETRIC VIEW

DRAW TOP VIEW

DRAW ISOMETRIC VIEW

THIRD ANGLE PROJECTION

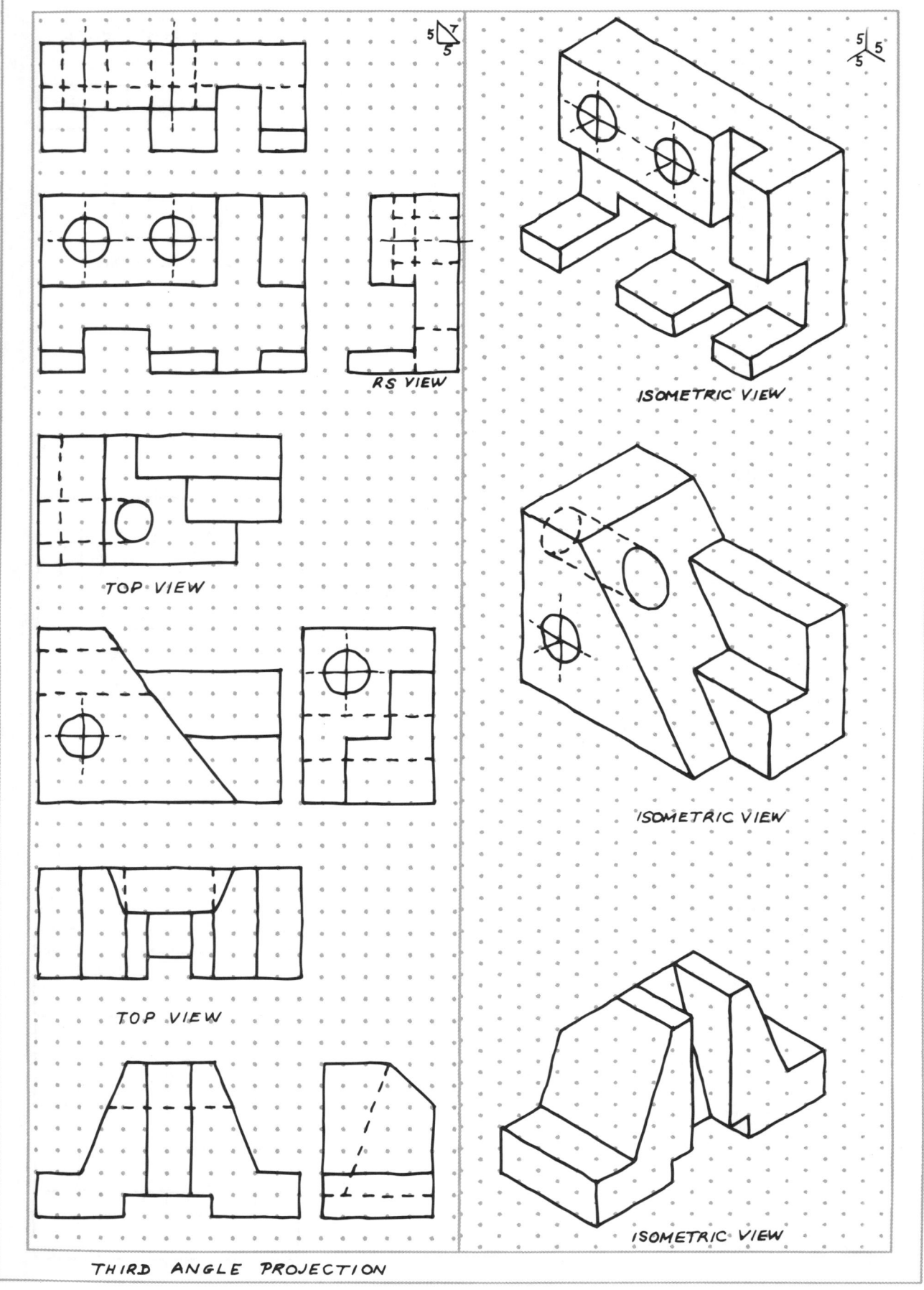

RS VIEW

ISOMETRIC VIEW

TOP VIEW

ISOMETRIC VIEW

TOP VIEW

ISOMETRIC VIEW

THIRD ANGLE PROJECTION

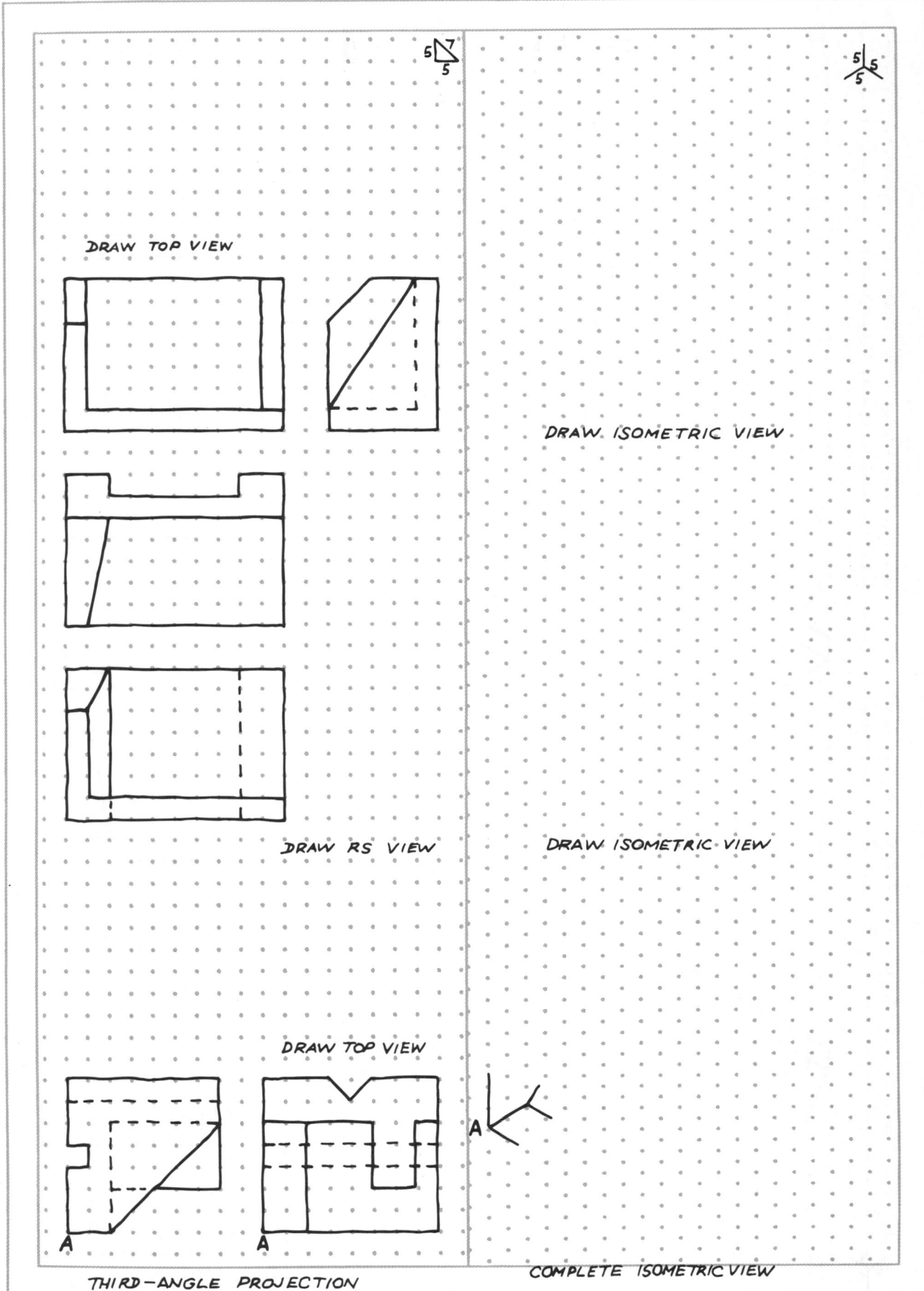

DRAW TOP VIEW

DRAW ISOMETRIC VIEW

DRAW RS VIEW

DRAW ISOMETRIC VIEW

DRAW TOP VIEW

A

THIRD—ANGLE PROJECTION

COMPLETE ISOMETRIC VIEW

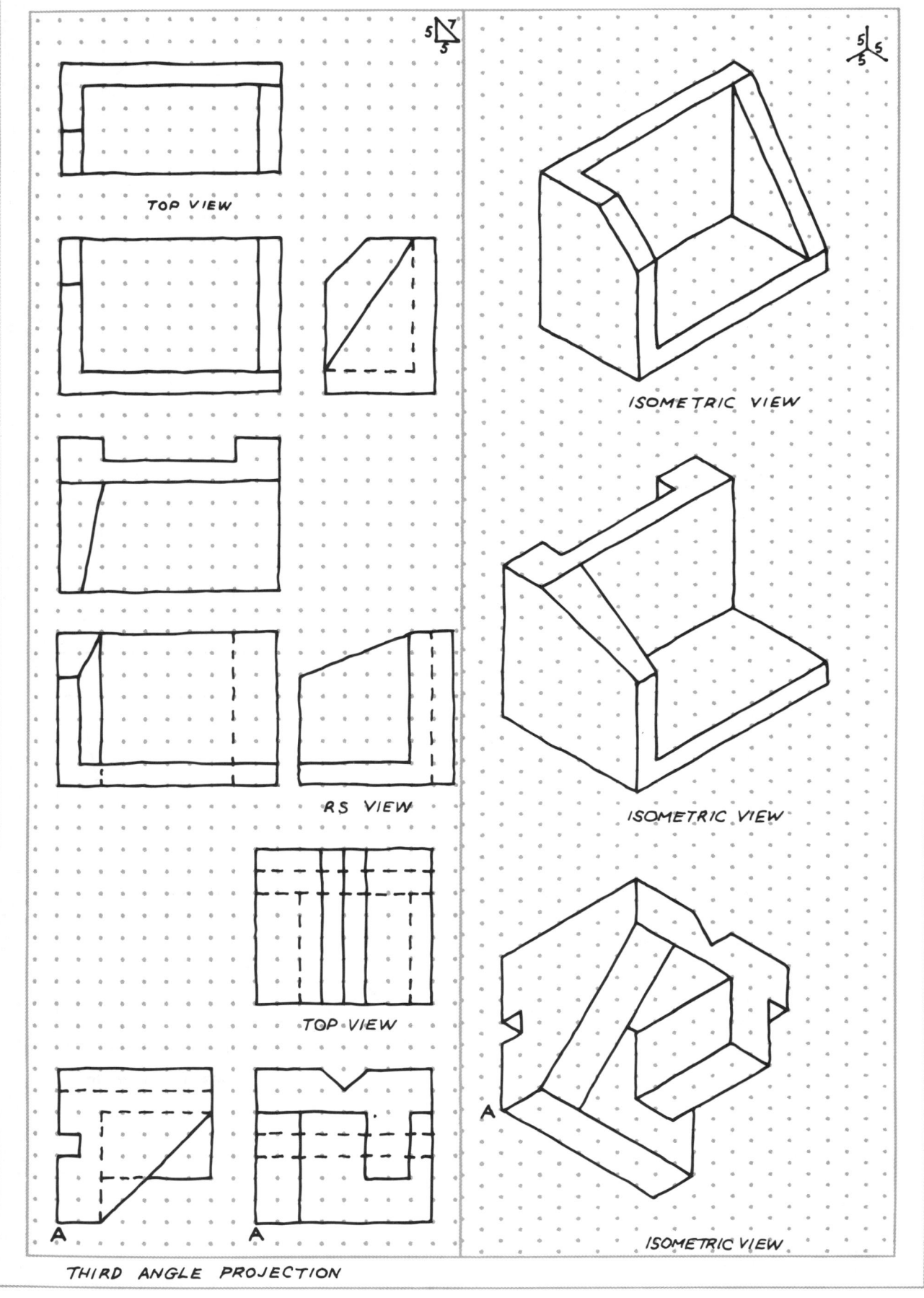

TOP VIEW

ISOMETRIC VIEW

RS VIEW

ISOMETRIC VIEW

TOP VIEW

A   A

THIRD ANGLE PROJECTION

ISOMETRIC VIEW

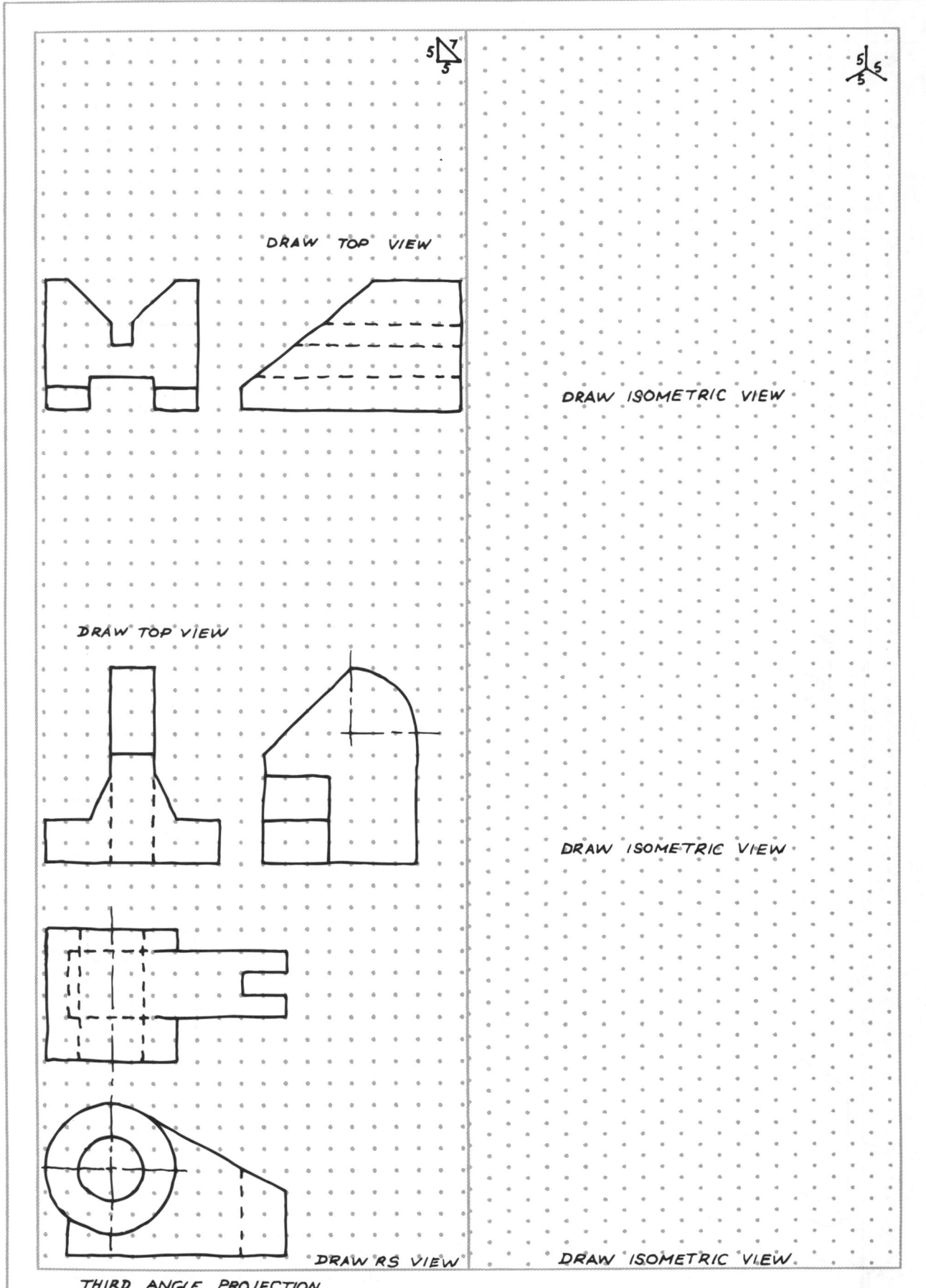

DRAW TOP VIEW

DRAW ISOMETRIC VIEW

DRAW TOP VIEW

DRAW ISOMETRIC VIEW

DRAW RS VIEW

DRAW ISOMETRIC VIEW

THIRD ANGLE PROJECTION

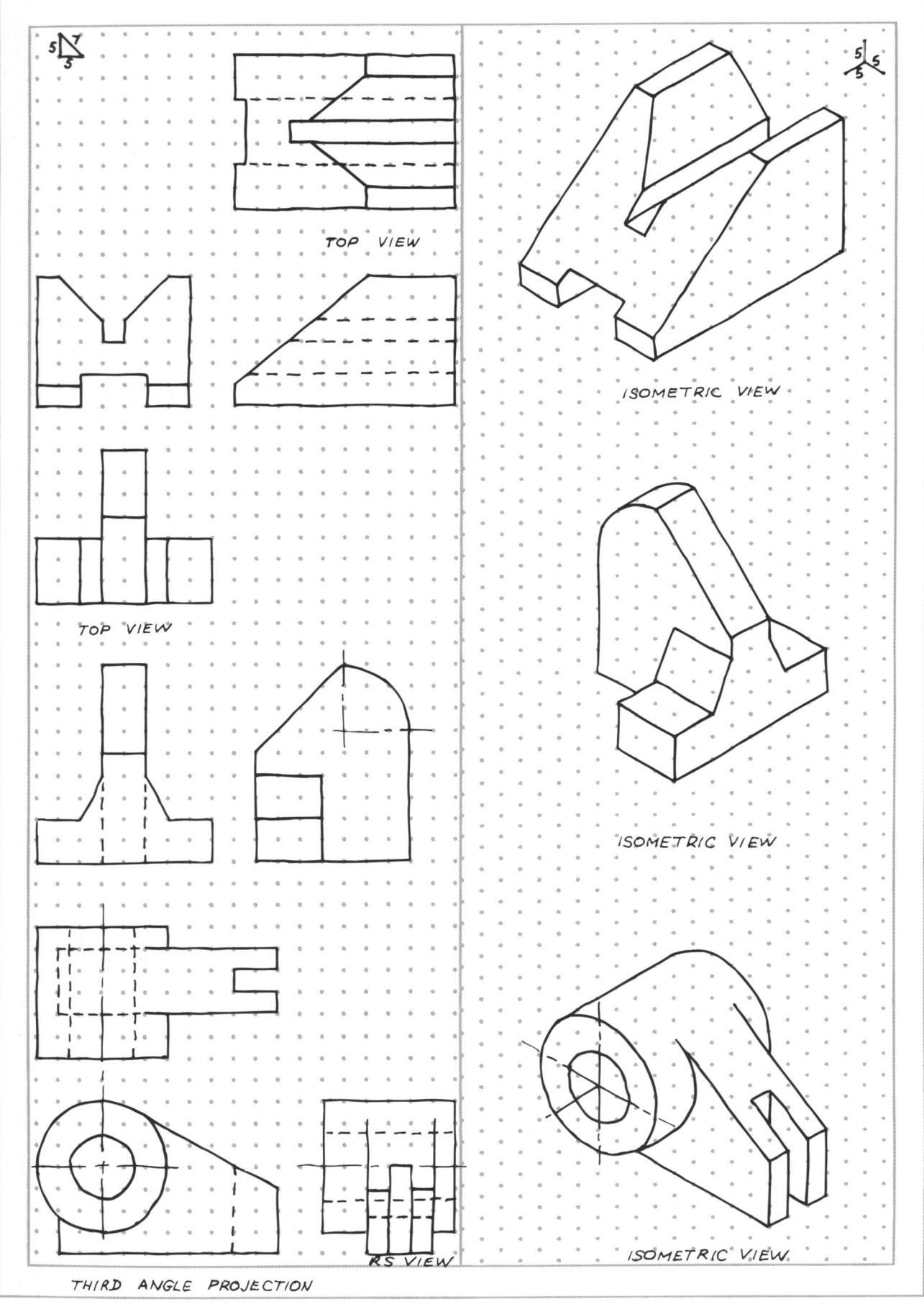

TOP VIEW

ISOMETRIC VIEW

TOP VIEW

ISOMETRIC VIEW

RS VIEW

THIRD ANGLE PROJECTION

ISOMETRIC VIEW

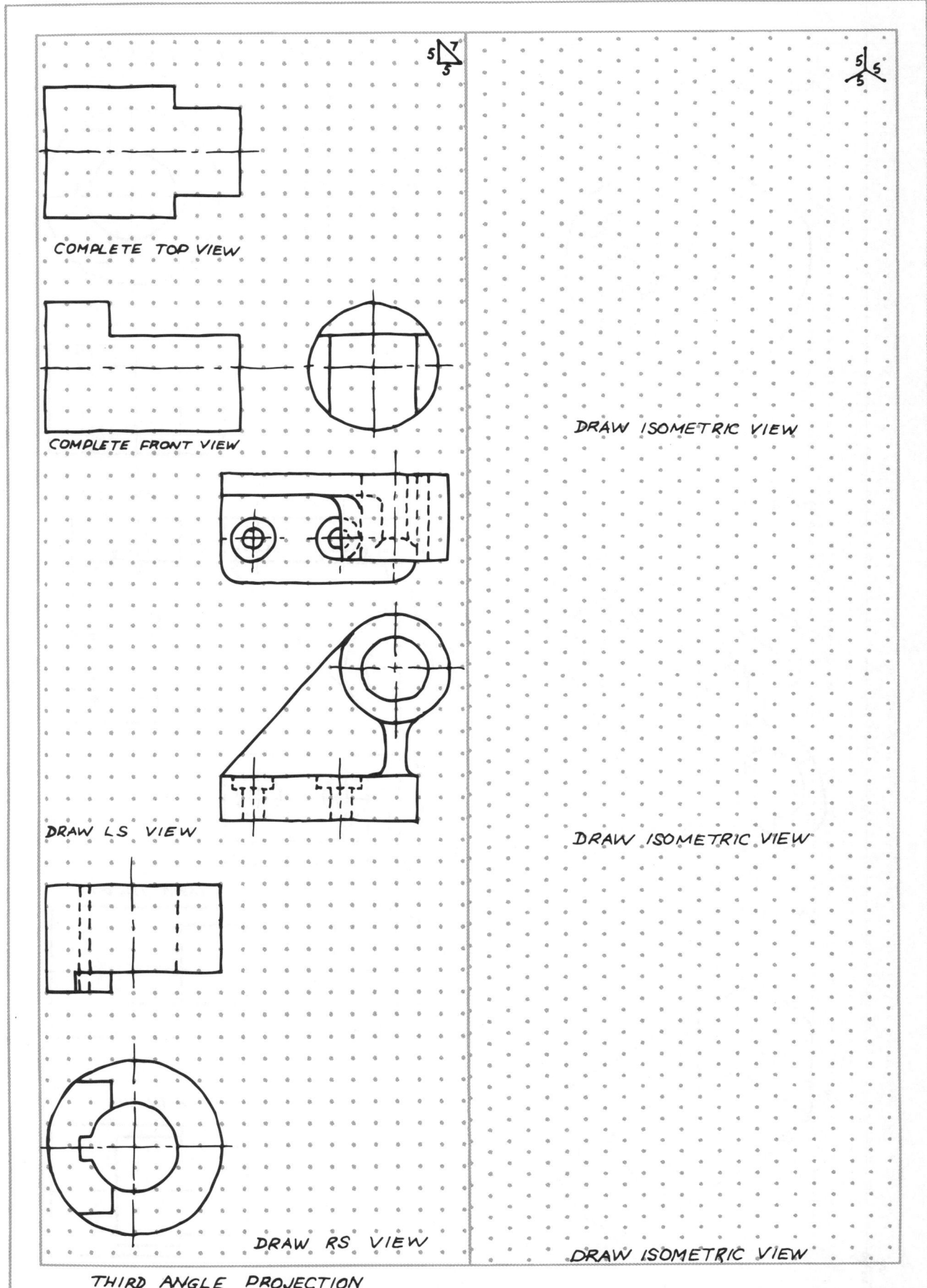

COMPLETE TOP VIEW

COMPLETE FRONT VIEW

DRAW ISOMETRIC VIEW

DRAW LS VIEW

DRAW ISOMETRIC VIEW

DRAW RS VIEW

DRAW ISOMETRIC VIEW

THIRD ANGLE PROJECTION

THIRD ANGLE PROJECTION

ISOMETRIC VIEW

RS VIEW

ISOMETRIC VIEW

LS VIEW

ISOMETRIC VIEW

FRONT VIEW

TOP VIEW

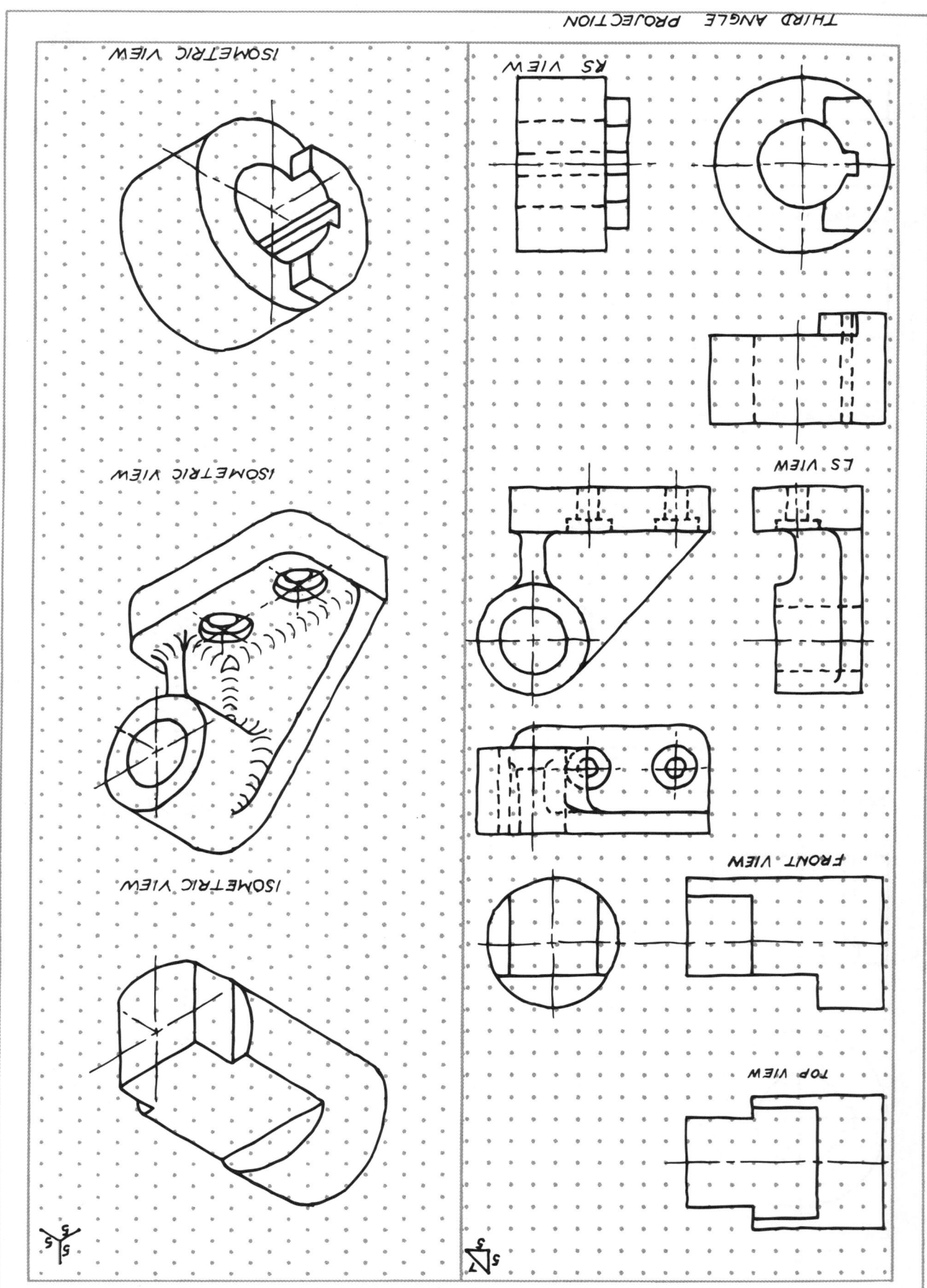

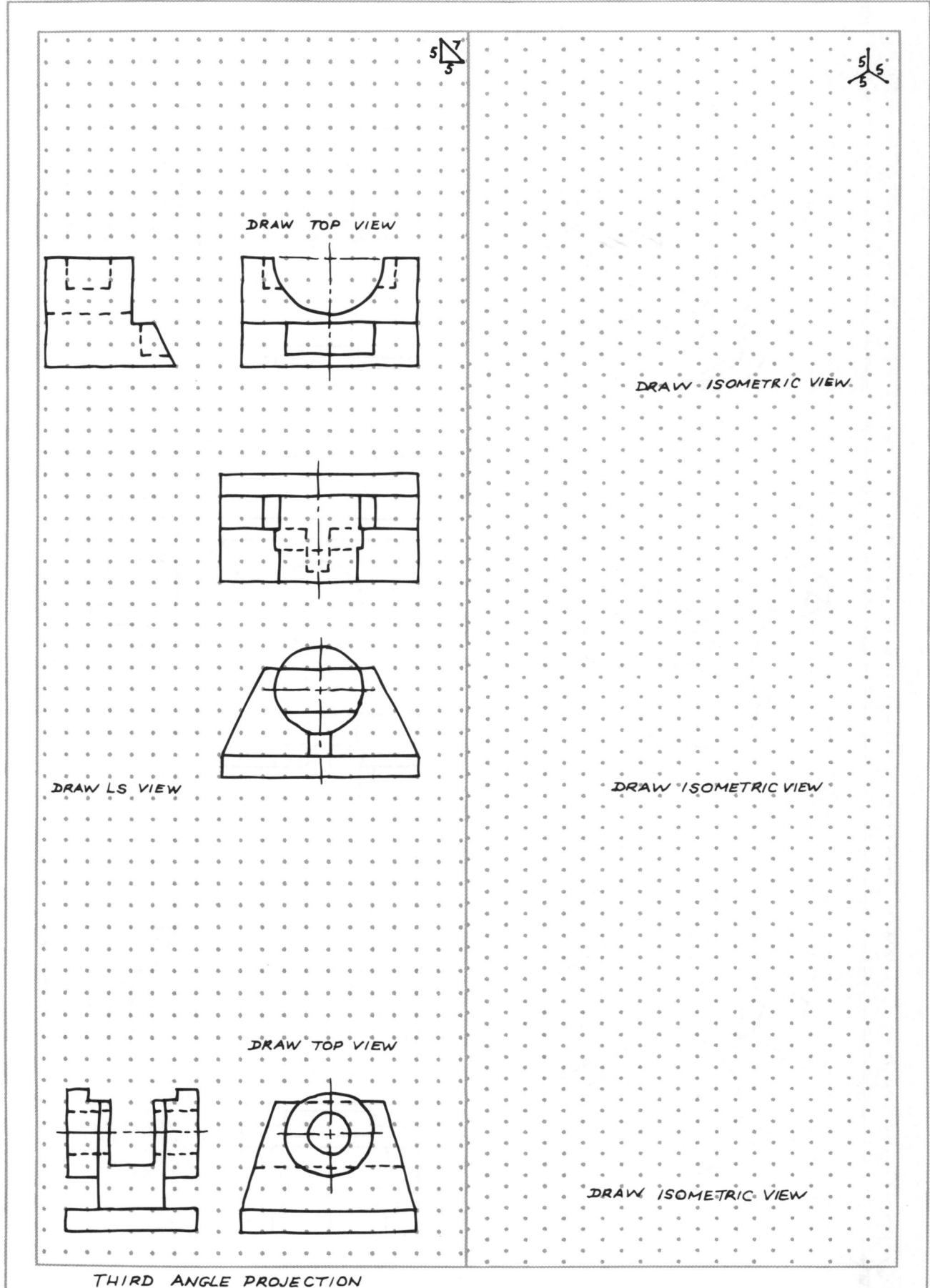

DRAW TOP VIEW

DRAW ISOMETRIC VIEW

DRAW LS VIEW

DRAW ISOMETRIC VIEW

DRAW TOP VIEW

DRAW ISOMETRIC VIEW

THIRD ANGLE PROJECTION

12

THIRD ANGLE PROJECTION

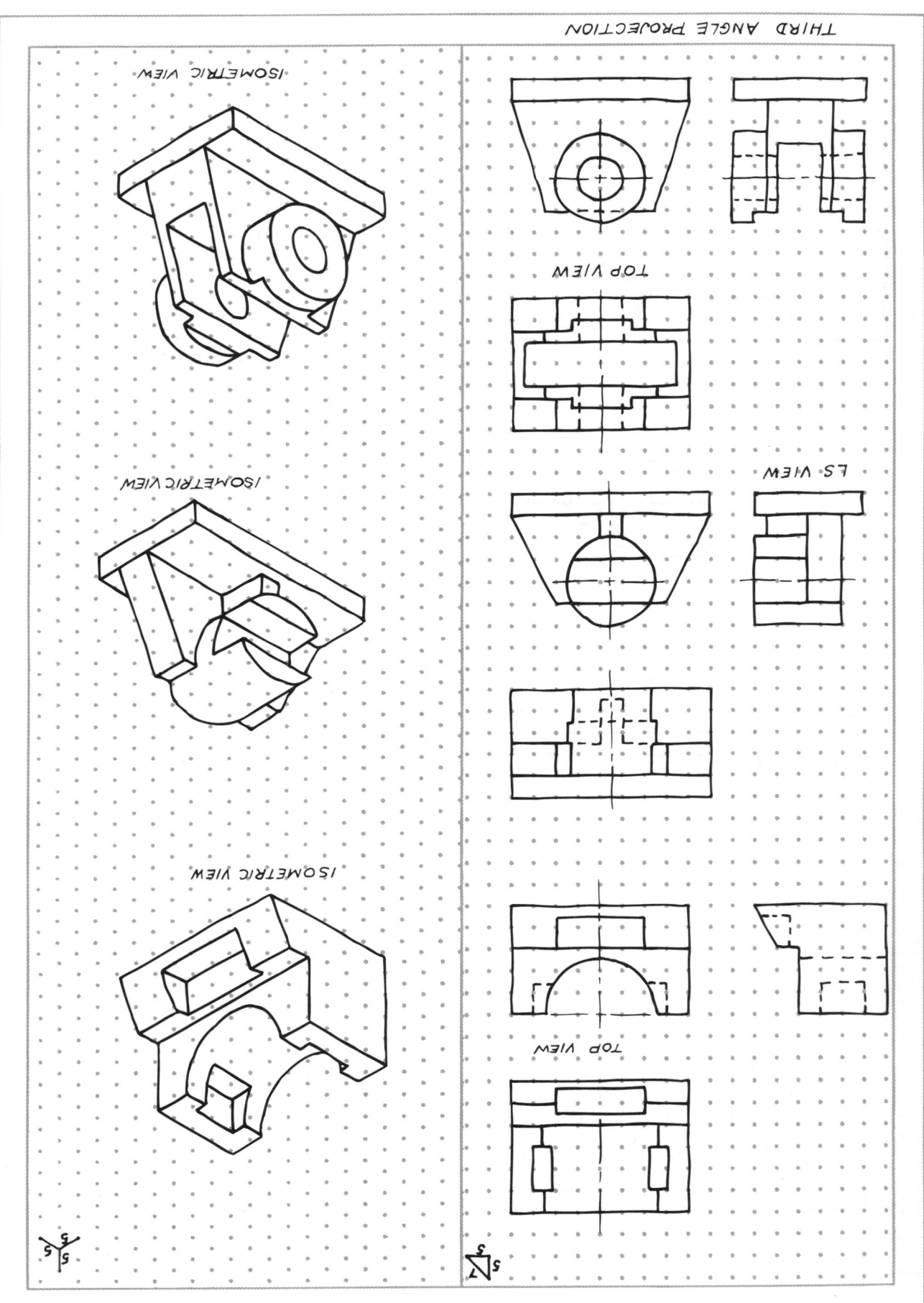

TOP VIEW

LS VIEW

ISOMETRIC VIEW

ISOMETRIC VIEW

TOP VIEW

ISOMETRIC VIEW

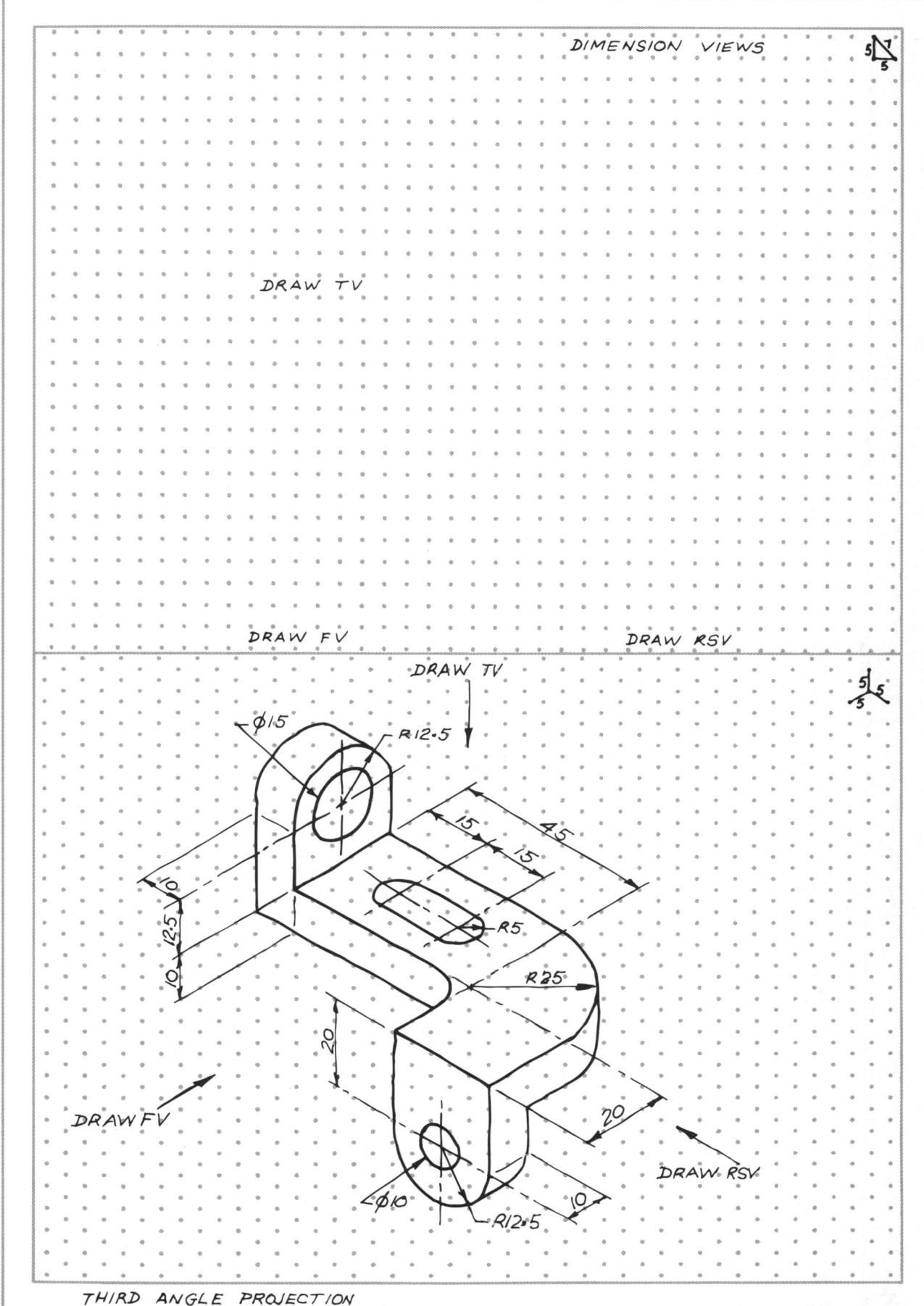

DRAW TV

DRAW FV   DRAW RSV

DRAW TV

Ø15   R12·5

15   15   45

R5

R25

10
12·5
10

20

20

DRAW FV

Ø10

R12·5   10

DRAW RSV

THIRD ANGLE PROJECTION

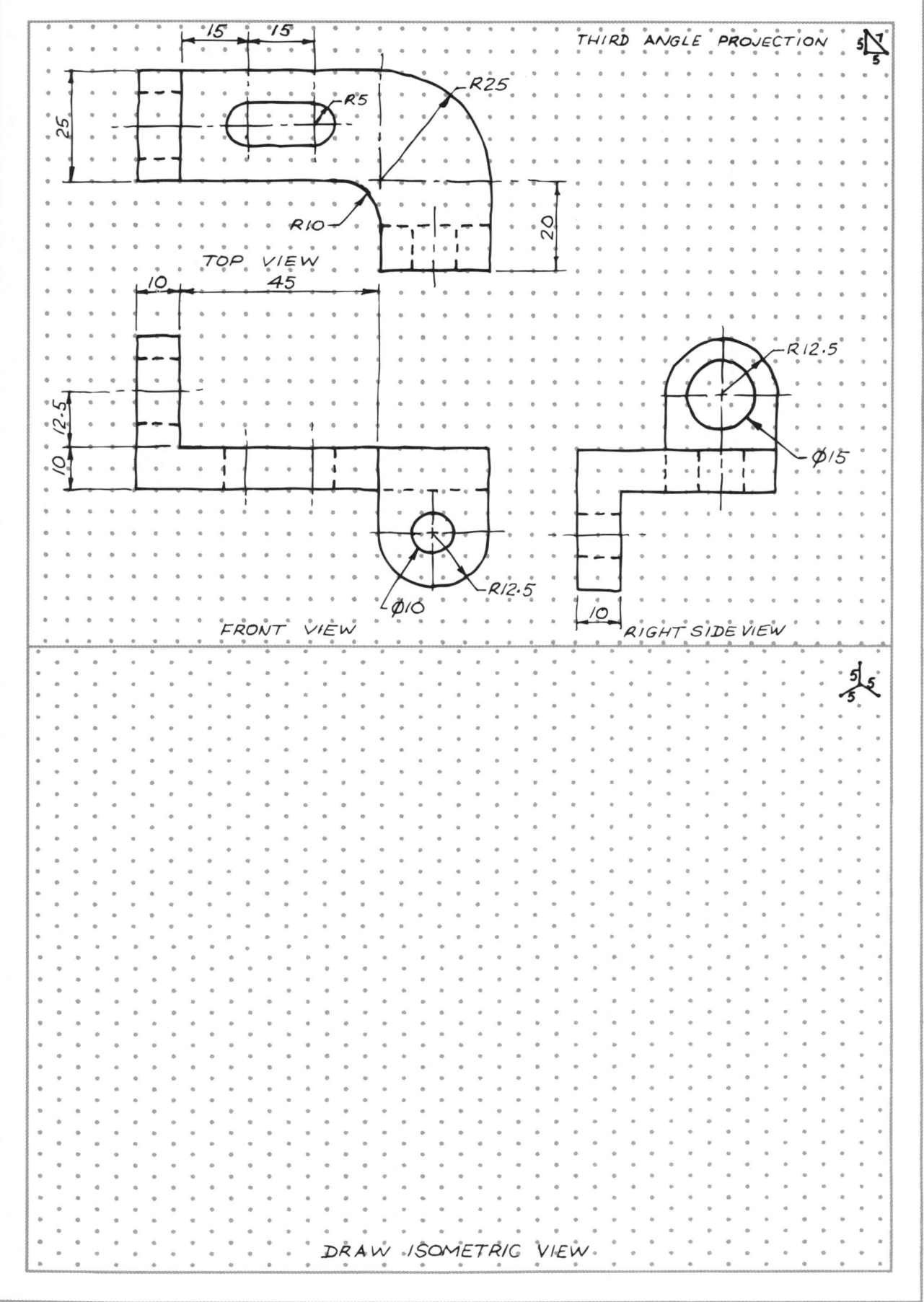

15 15

25

R5

R25

R10

TOP VIEW

20

10

45

12.5

10

R12.5

Ø15

Ø10

R12.5

FRONT VIEW

RIGHT SIDE VIEW

10

DRAW ISOMETRIC VIEW

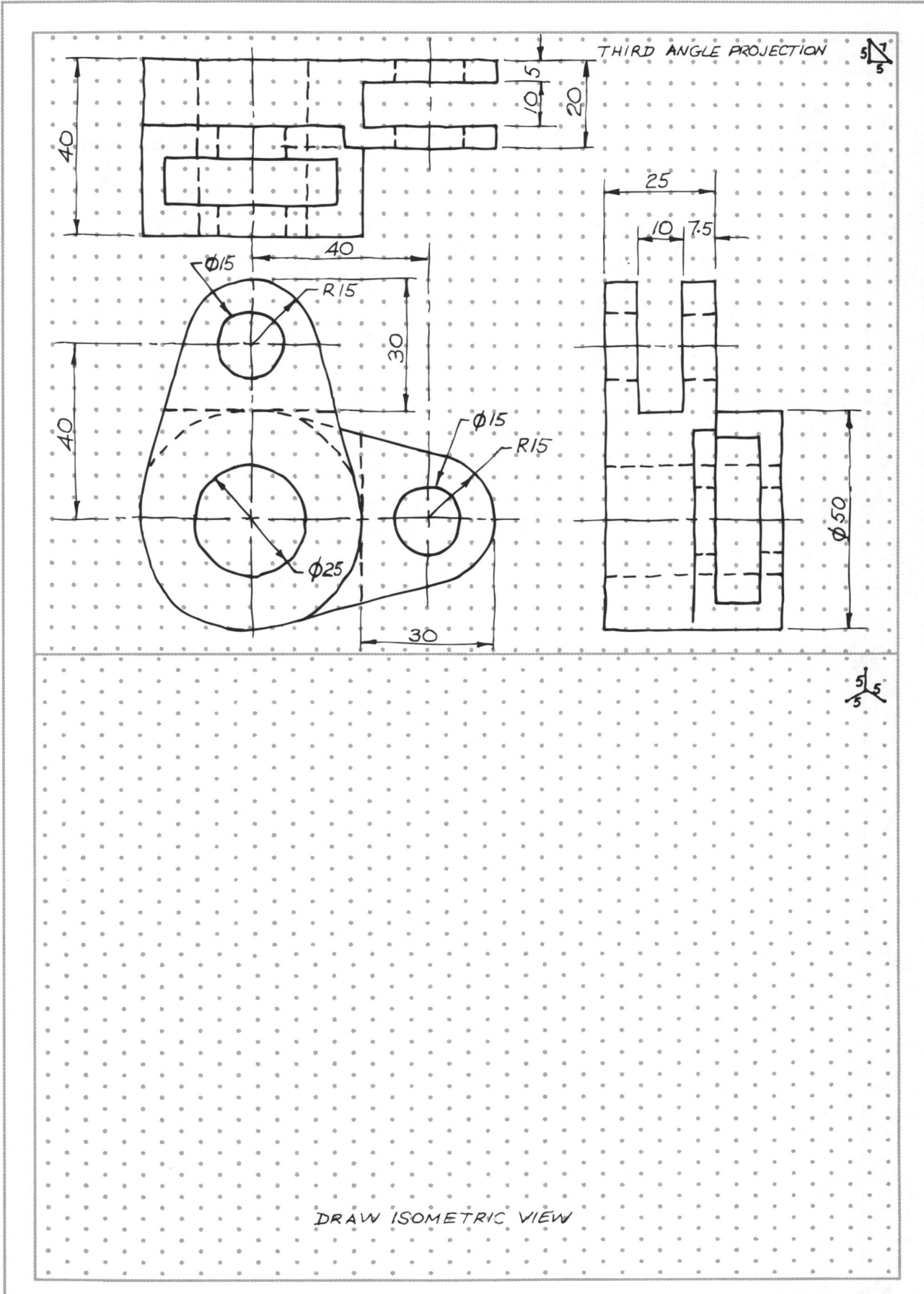

THIRD ANGLE PROJECTION

Ø15
R15
40
30
Ø15
R15
Ø25
30

10·5
20
40

25
10 7·5
Ø50

DRAW ISOMETRIC VIEW

15

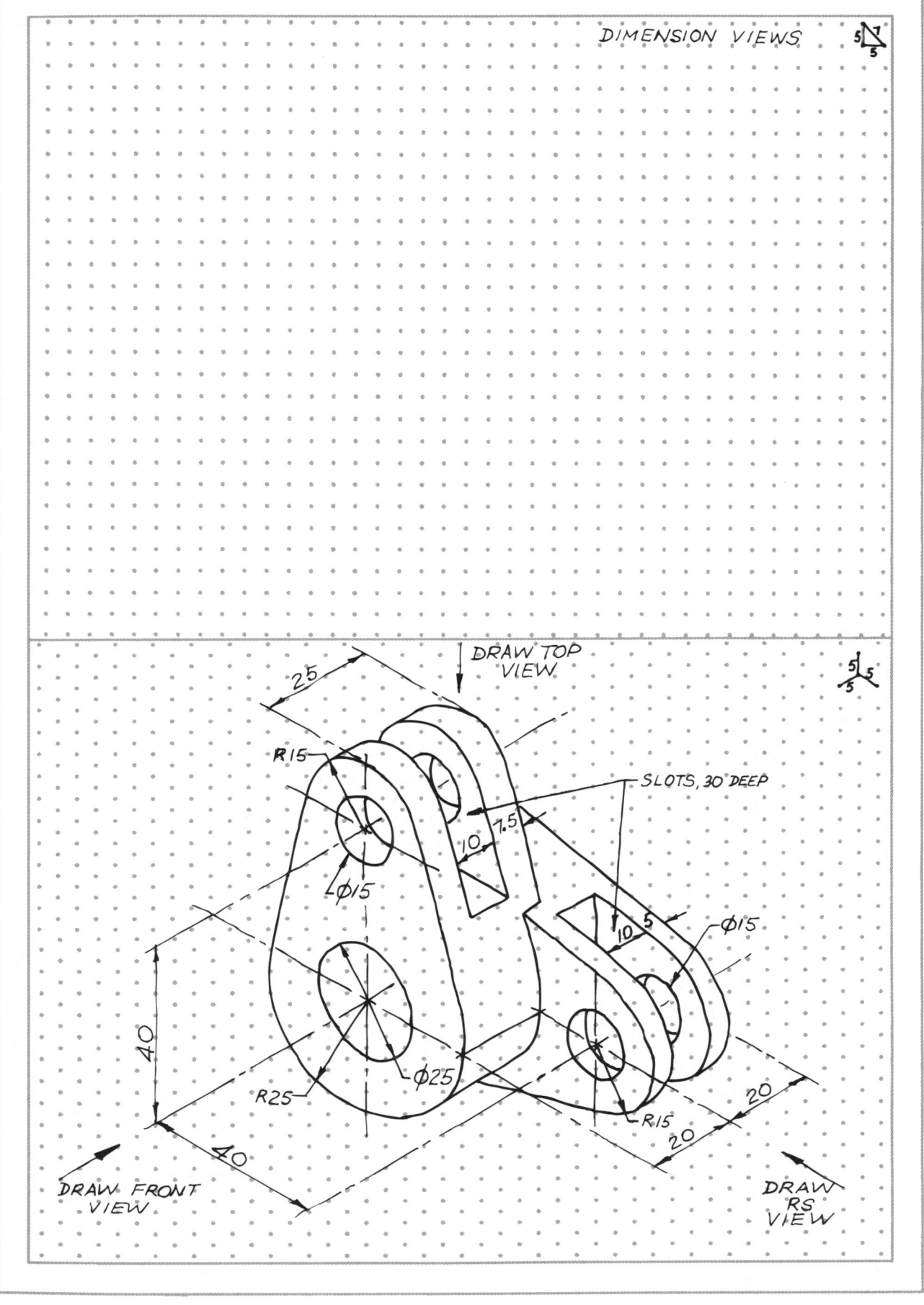

DRAW TOP
VIEW

25

R15

SLOTS, 30° DEEP

10   7.5

Ø15

10   5

Ø15

40

R25

Ø25

R15

20

20

40

DRAW FRONT
VIEW

DRAW
RS
VIEW

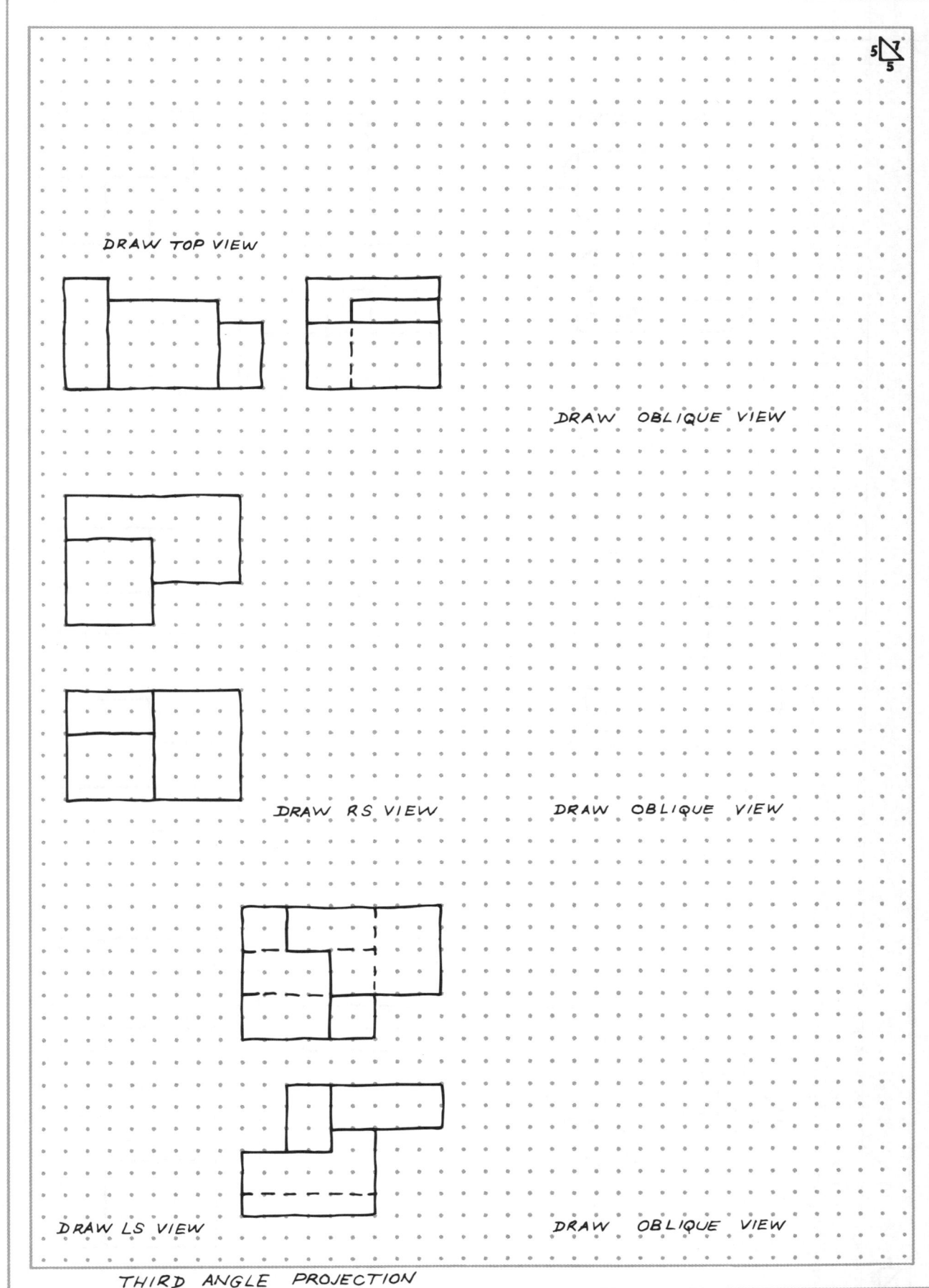

DRAW TOP VIEW

DRAW OBLIQUE VIEW

DRAW RS VIEW

DRAW OBLIQUE VIEW

DRAW LS VIEW

DRAW OBLIQUE VIEW

THIRD ANGLE PROJECTION

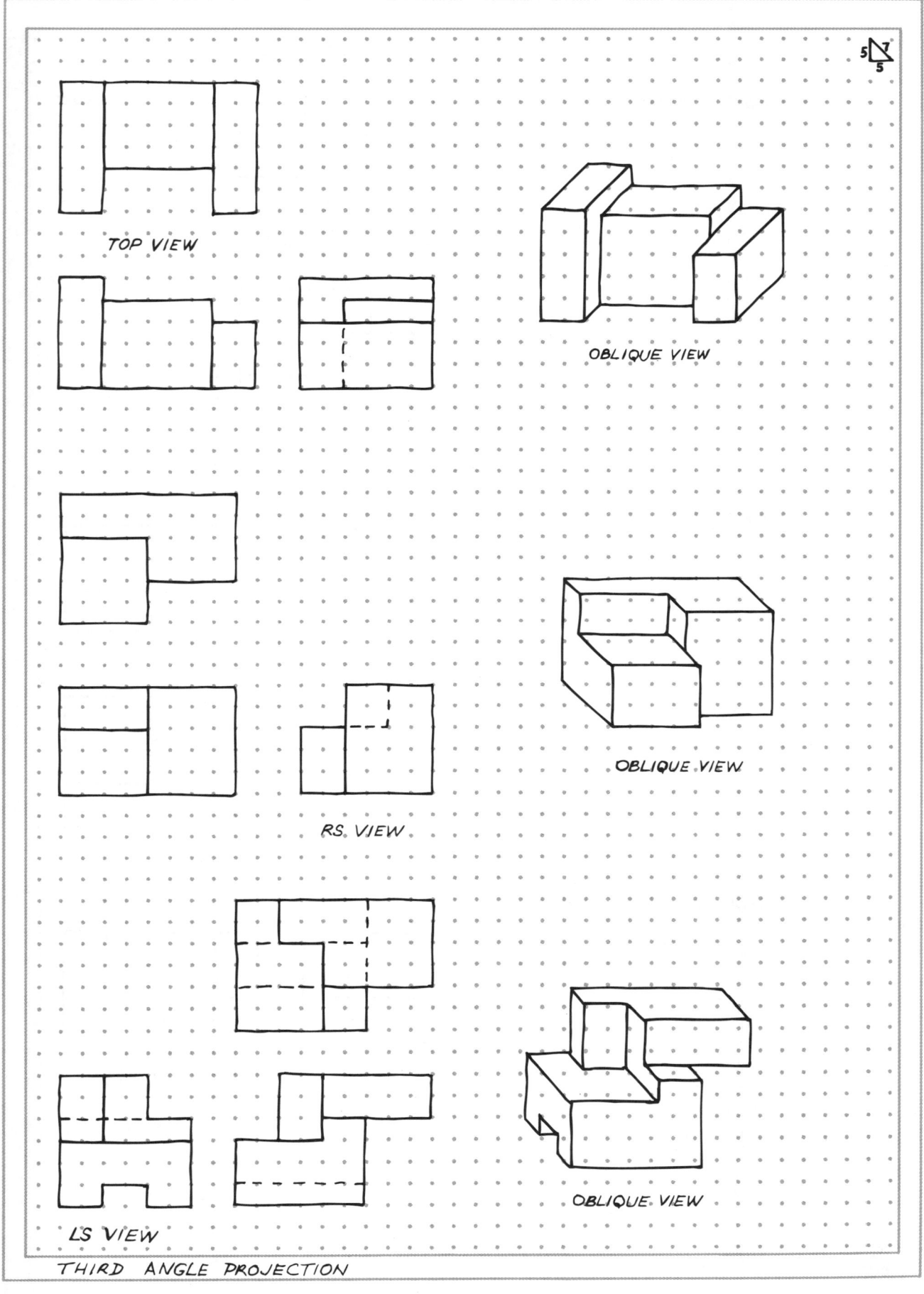

TOP VIEW

OBLIQUE VIEW

RS. VIEW

OBLIQUE VIEW

LS VIEW

OBLIQUE VIEW

THIRD ANGLE PROJECTION

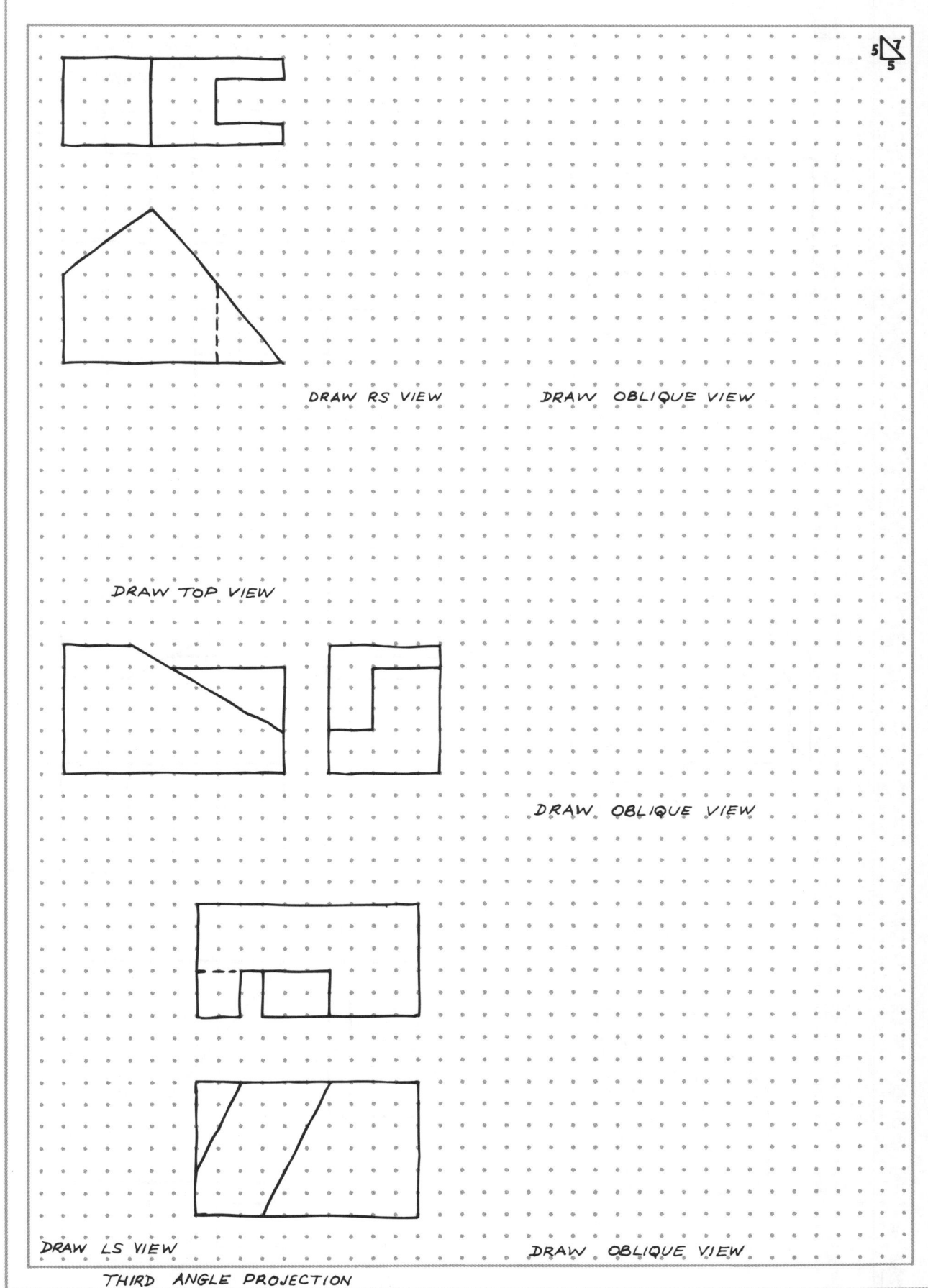

DRAW RS VIEW   DRAW OBLIQUE VIEW

DRAW TOP VIEW

DRAW OBLIQUE VIEW

DRAW LS VIEW   DRAW OBLIQUE VIEW

THIRD ANGLE PROJECTION

THIRD ANGLE PROJECTION

LS VIEW

OBLIQUE VIEW

OBLIQUE VIEW

TOP VIEW

RS VIEW

OBLIQUE VIEW

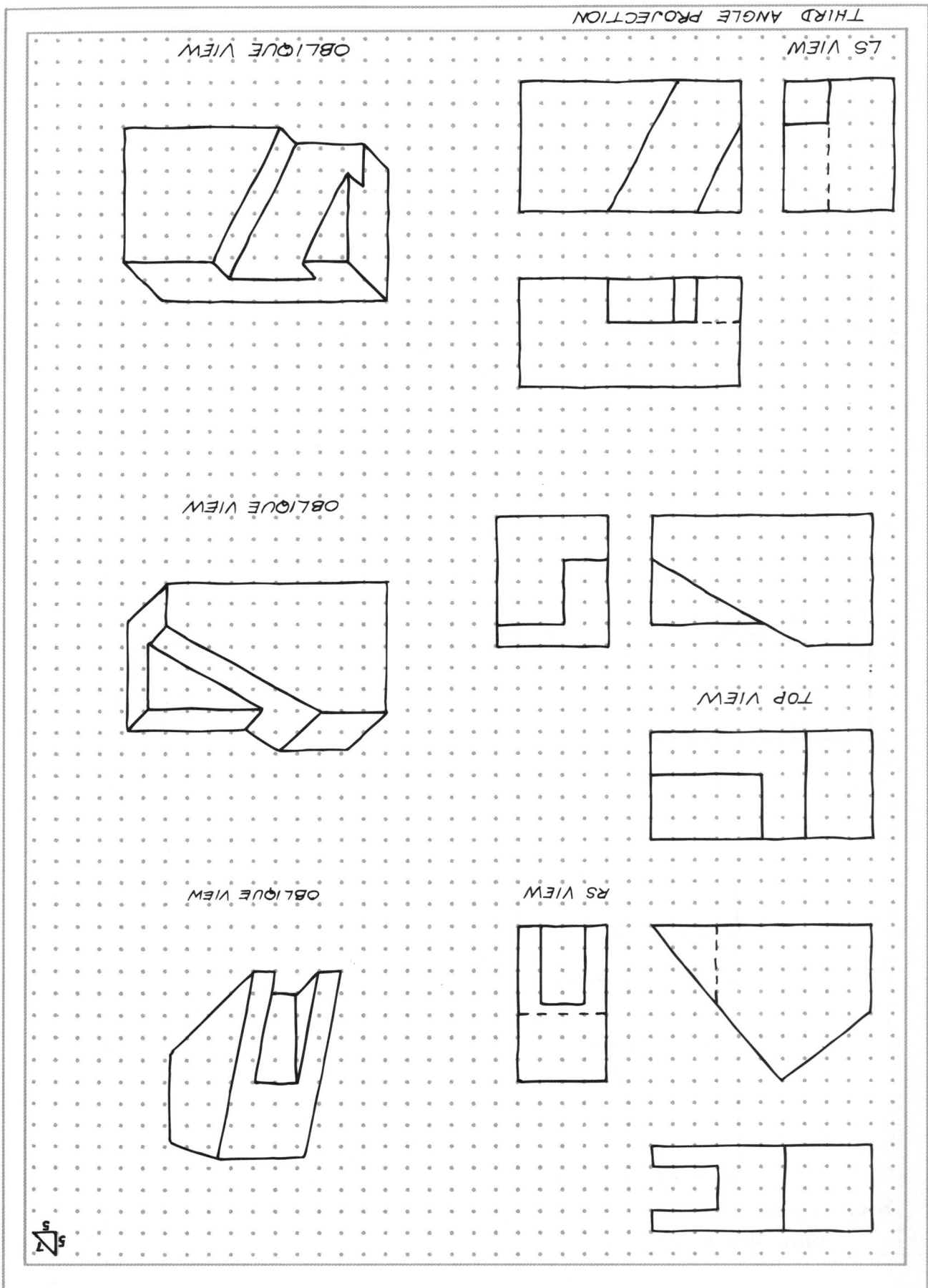

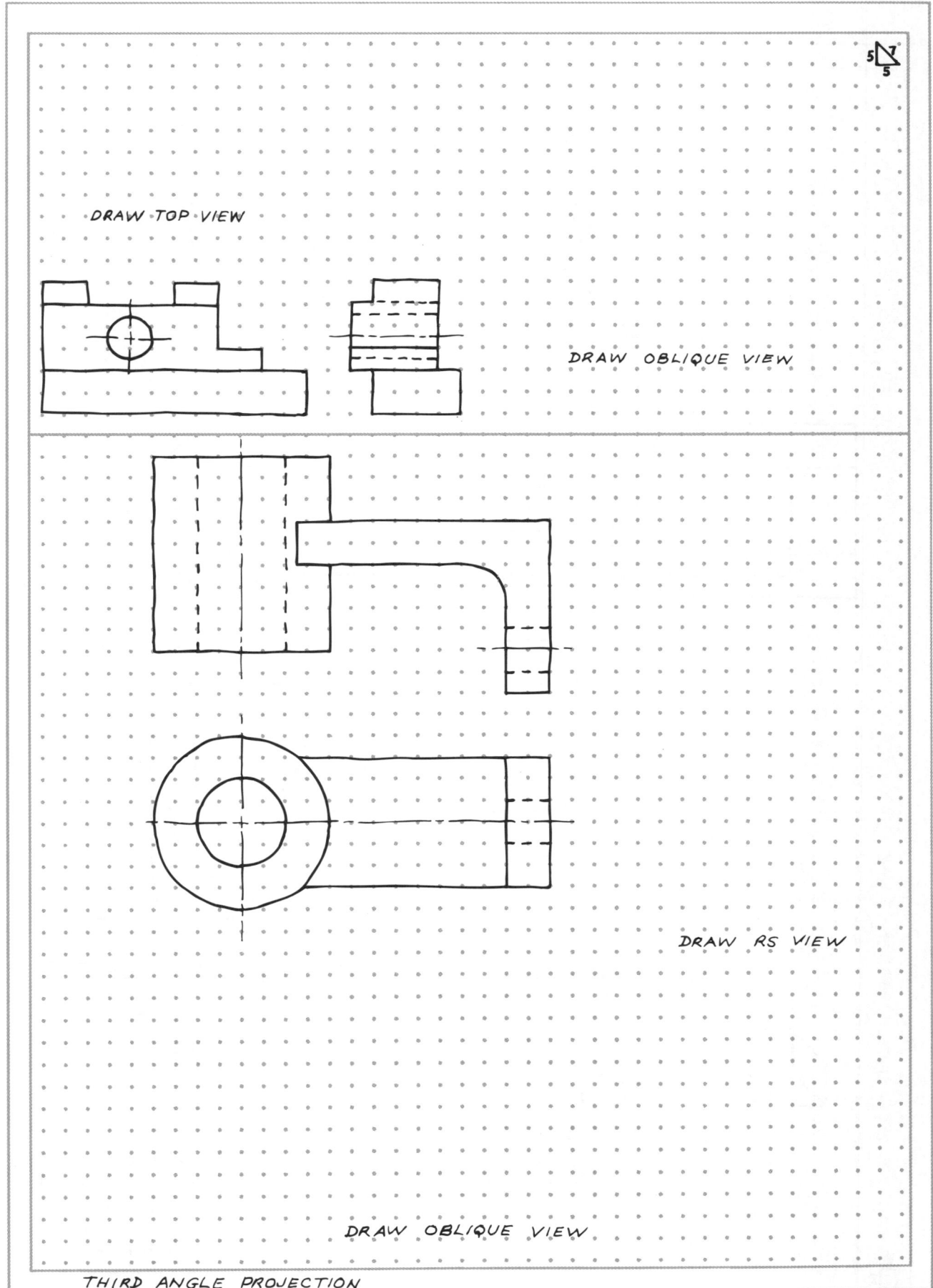

DRAW TOP VIEW

DRAW OBLIQUE VIEW

DRAW RS VIEW

DRAW OBLIQUE VIEW

THIRD ANGLE PROJECTION

THIRD ANGLE PROJECTION

OBLIQUE VIEW

RS VIEW

OBLIQUE VIEW

TOP VIEW

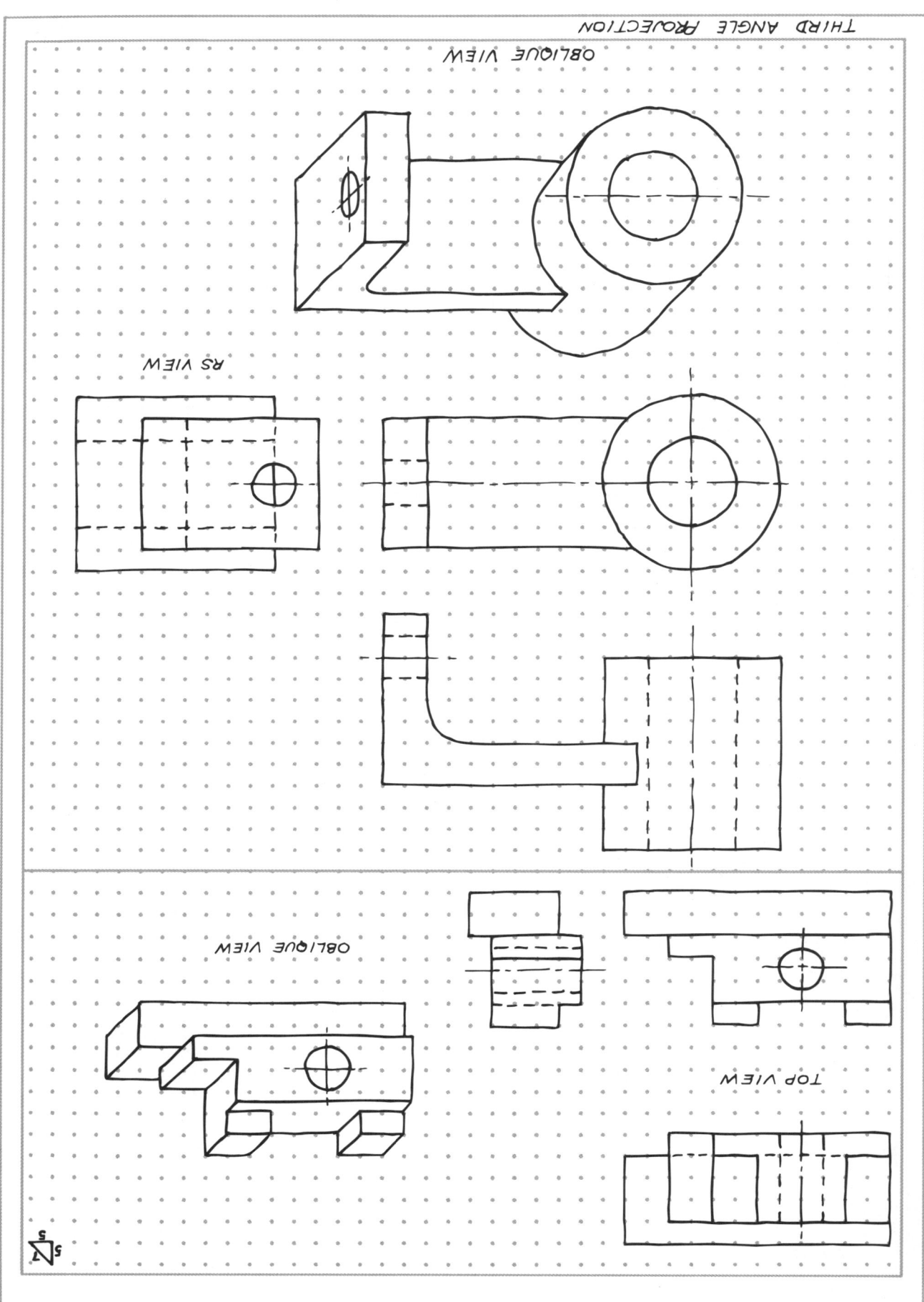

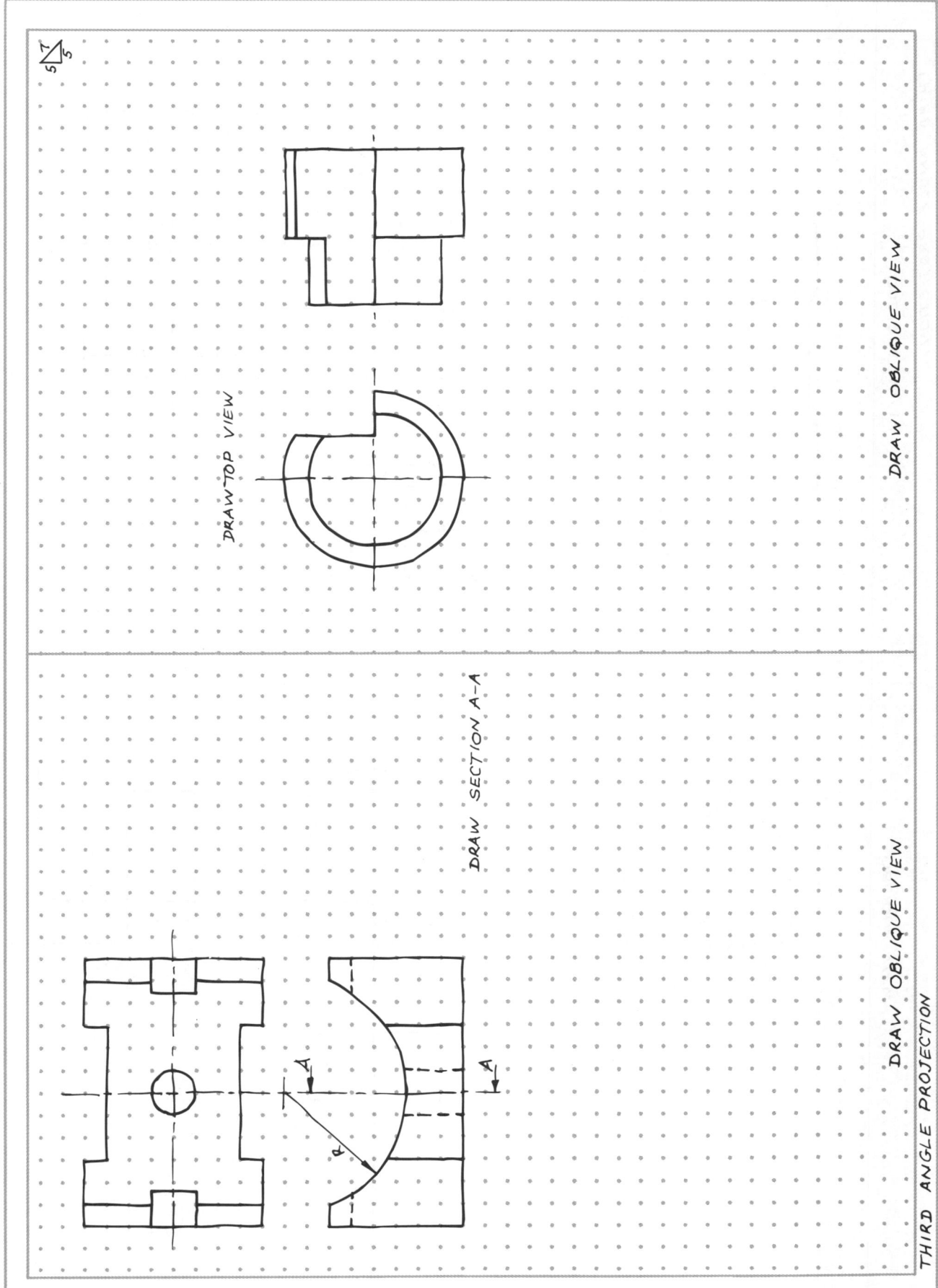

DRAW TOP VIEW

DRAW OBLIQUE VIEW

DRAW SECTION A-A

DRAW OBLIQUE VIEW

THIRD ANGLE PROJECTION

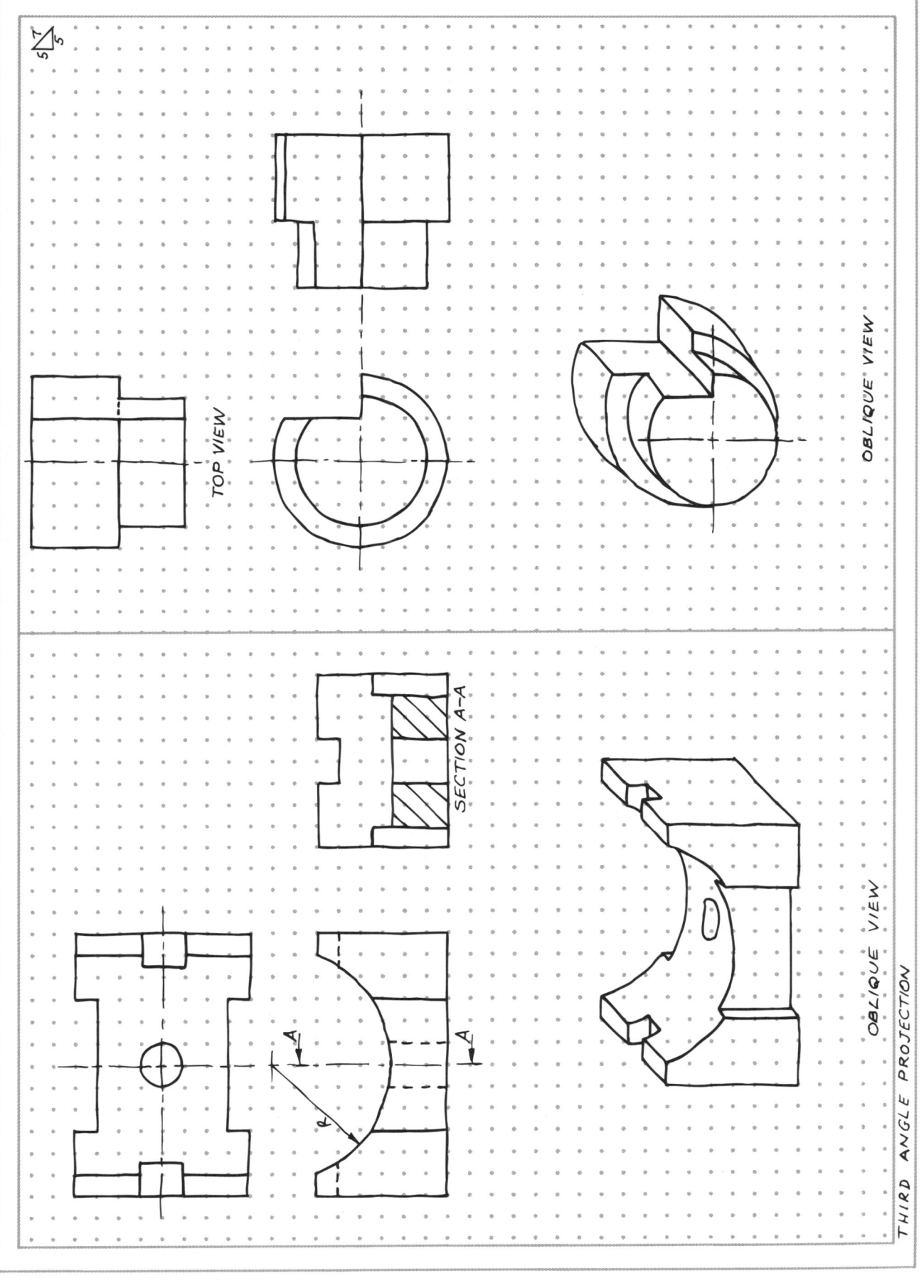

TOP VIEW

OBLIQUE VIEW

SECTION A-A

OBLIQUE VIEW

THIRD ANGLE PROJECTION

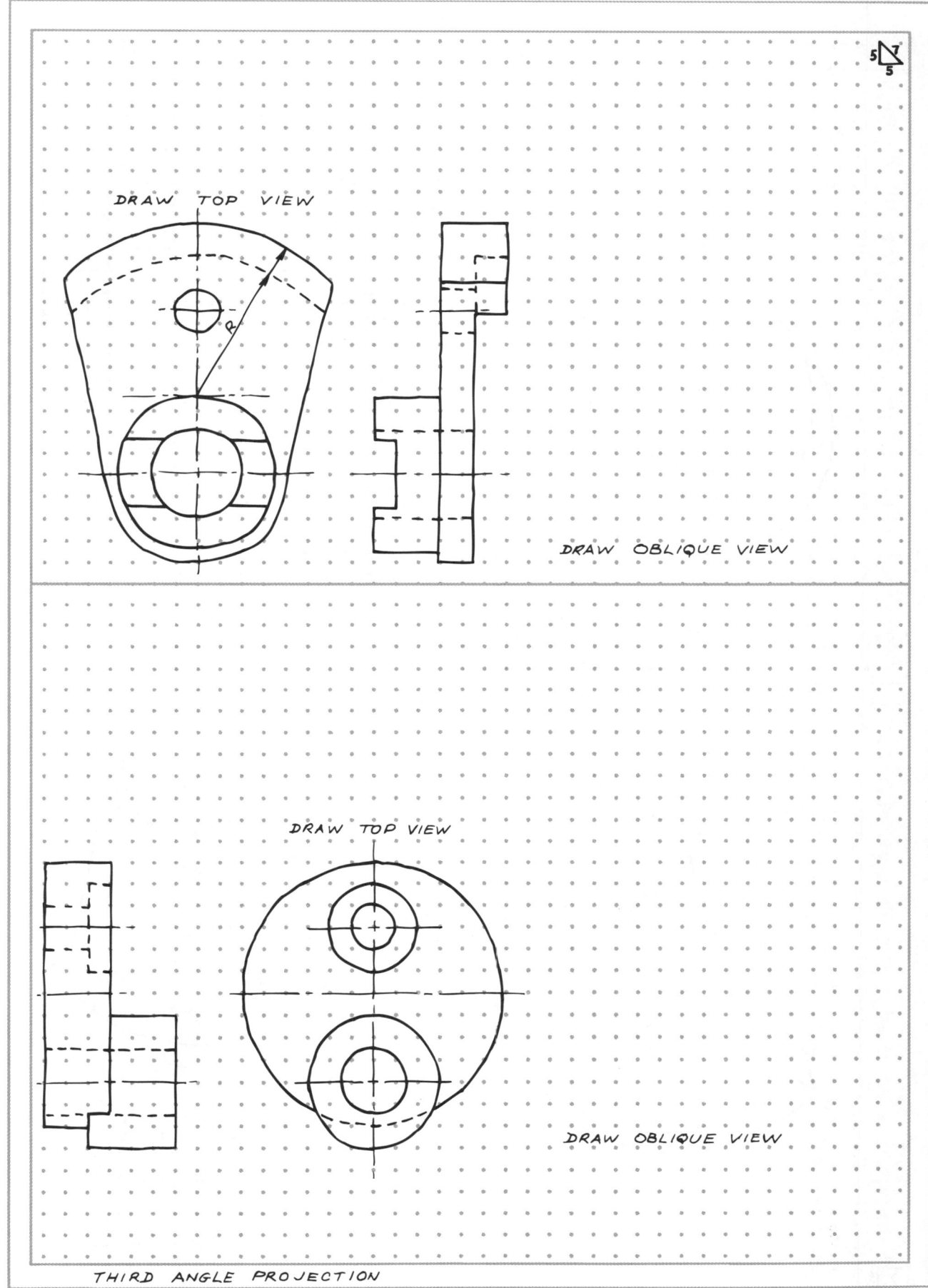

DRAW TOP VIEW

DRAW OBLIQUE VIEW

DRAW TOP VIEW

DRAW OBLIQUE VIEW

THIRD ANGLE PROJECTION

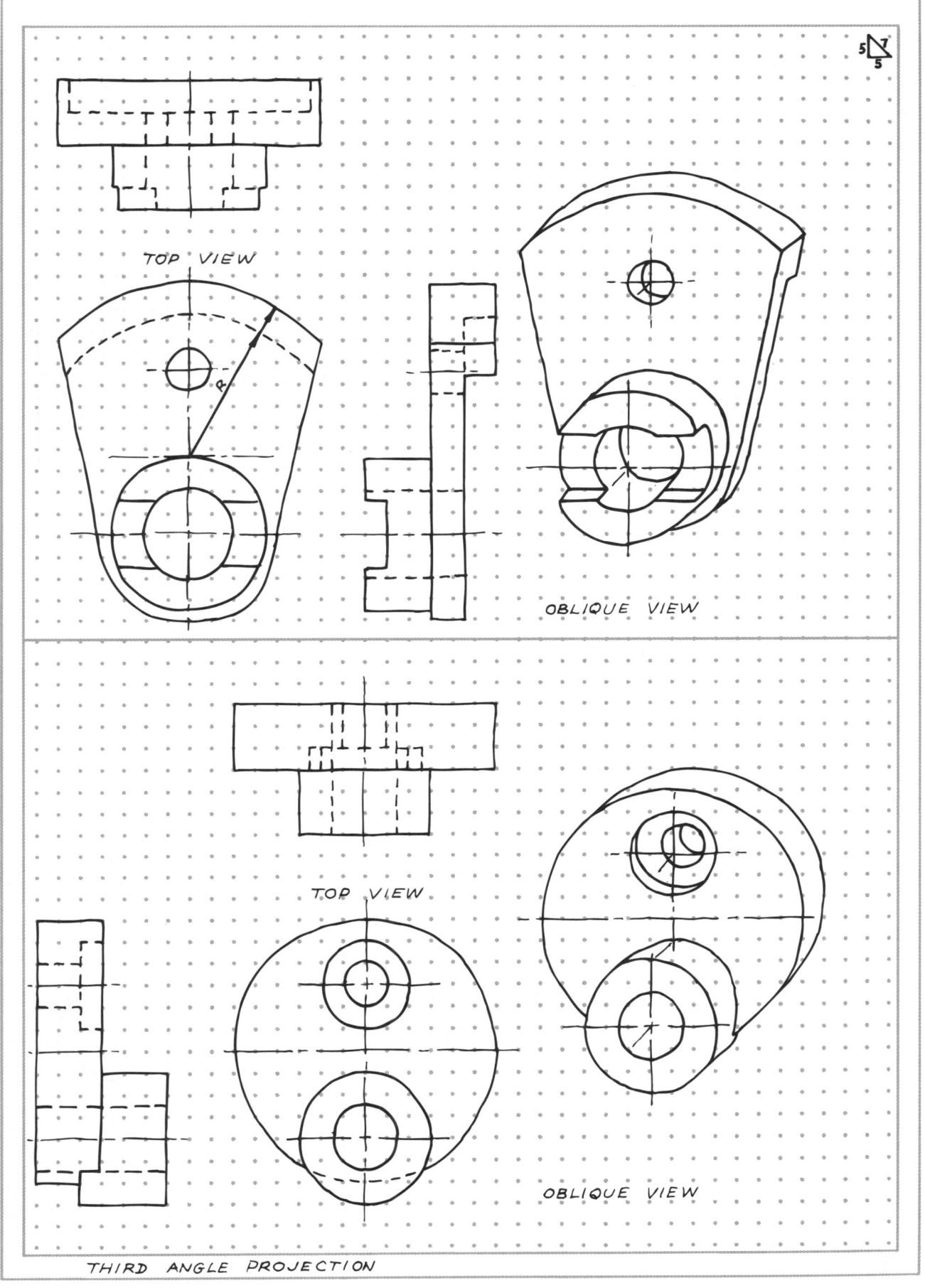

TOP VIEW

OBLIQUE VIEW

TOP VIEW

OBLIQUE VIEW

THIRD ANGLE PROJECTION

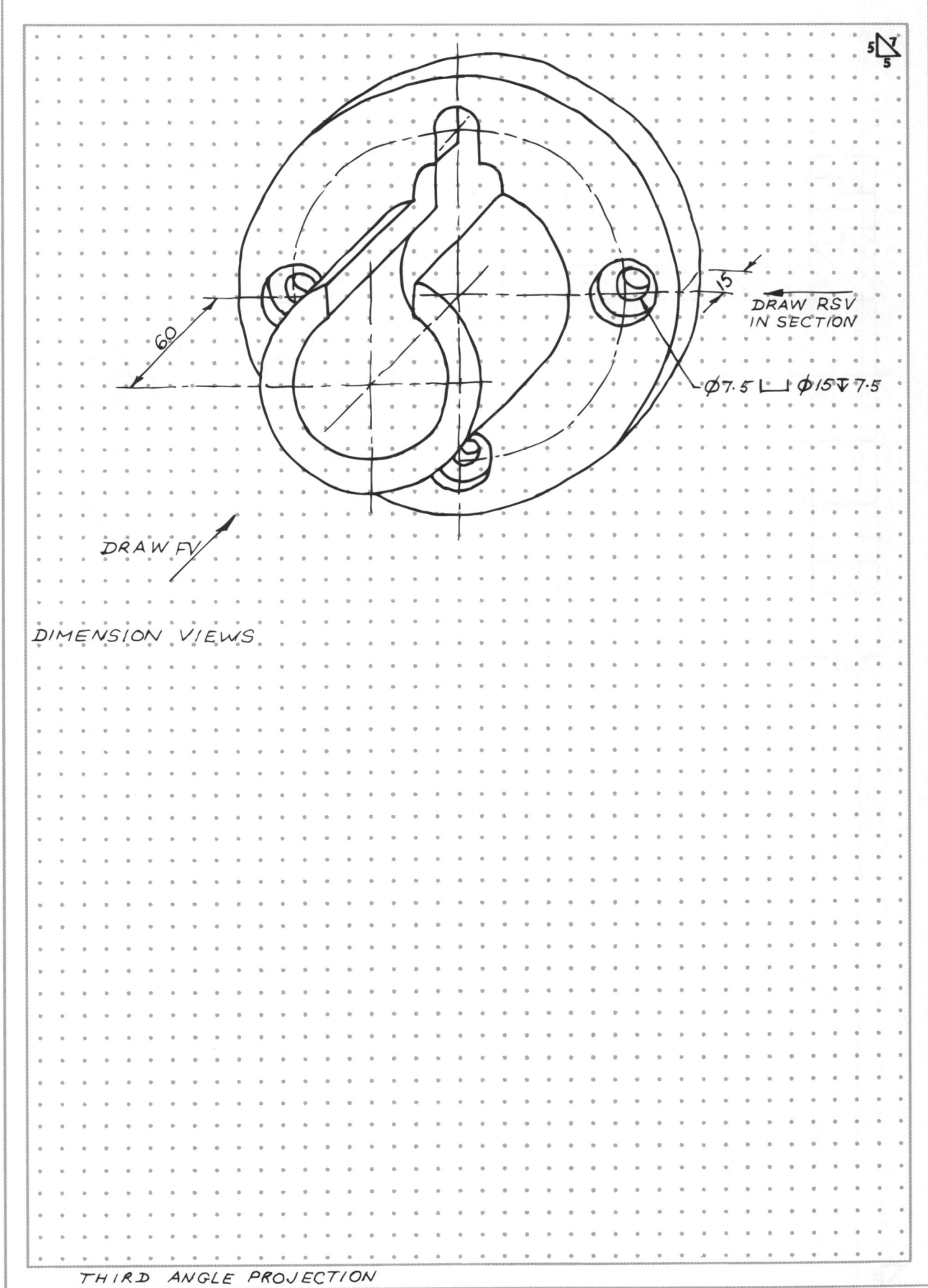

DRAW RSV
IN SECTION

60

15

∅7.5 ⌴ ∅15 ⫂ 7.5

DRAW FV

DIMENSION VIEWS

5 ◿ 7
  5

THIRD ANGLE PROJECTION

27

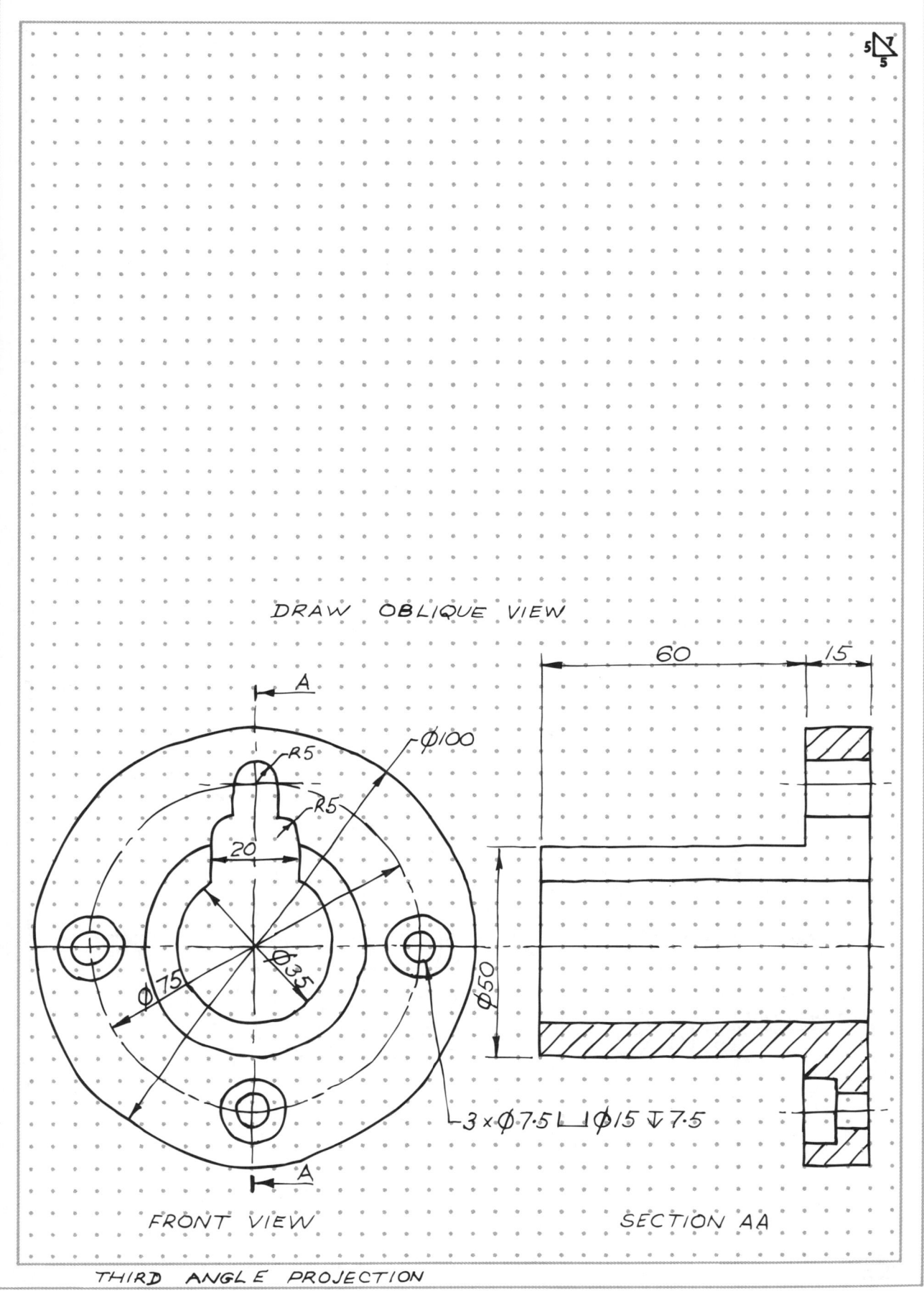

DRAW OBLIQUE VIEW

A

φ100

R5

R5

20

φ75

φ35

60

15

φ50

3 × φ7·5 ⌴ φ15 ⌄ 7·5

A

FRONT VIEW

SECTION AA

THIRD ANGLE PROJECTION

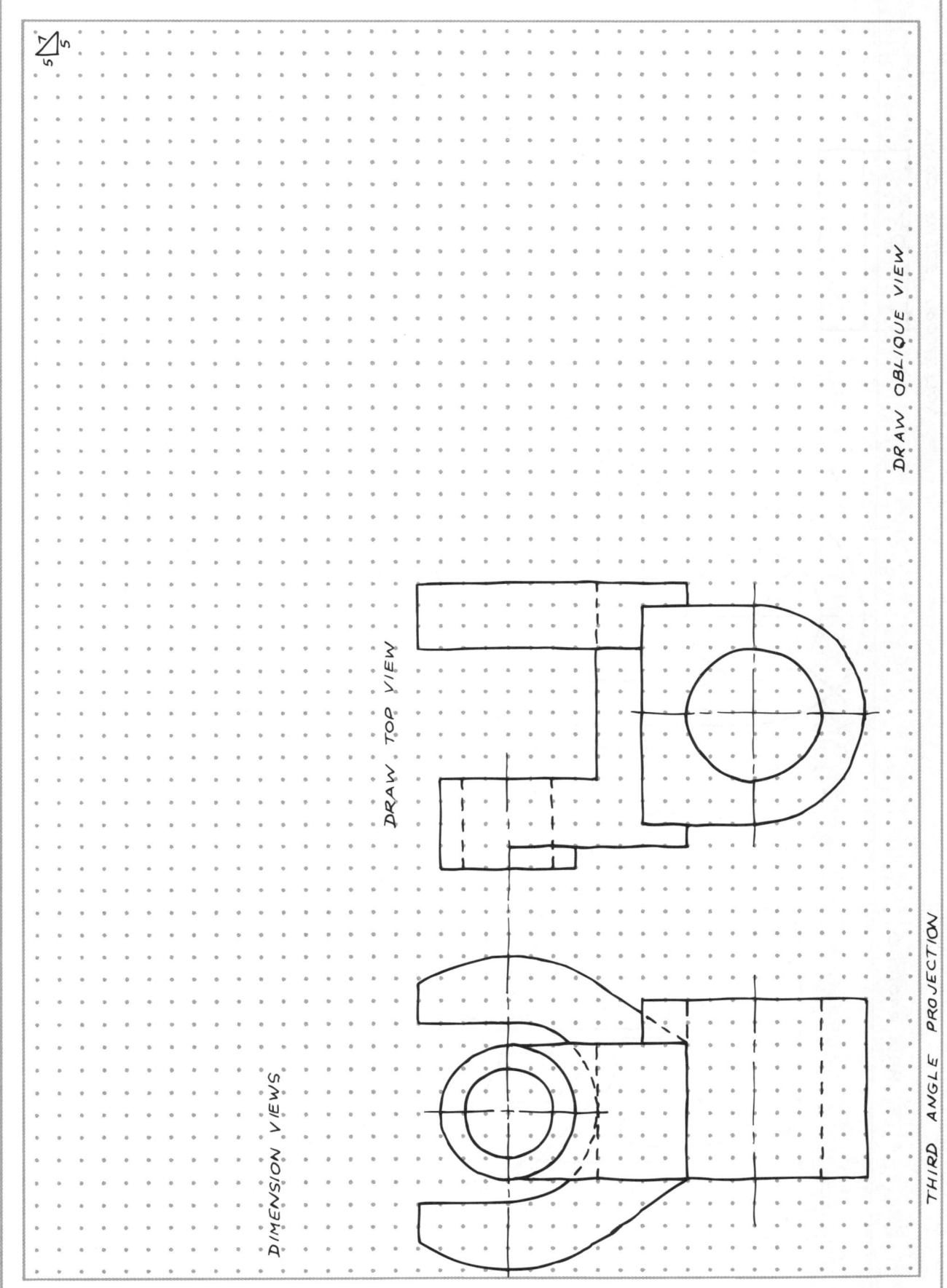

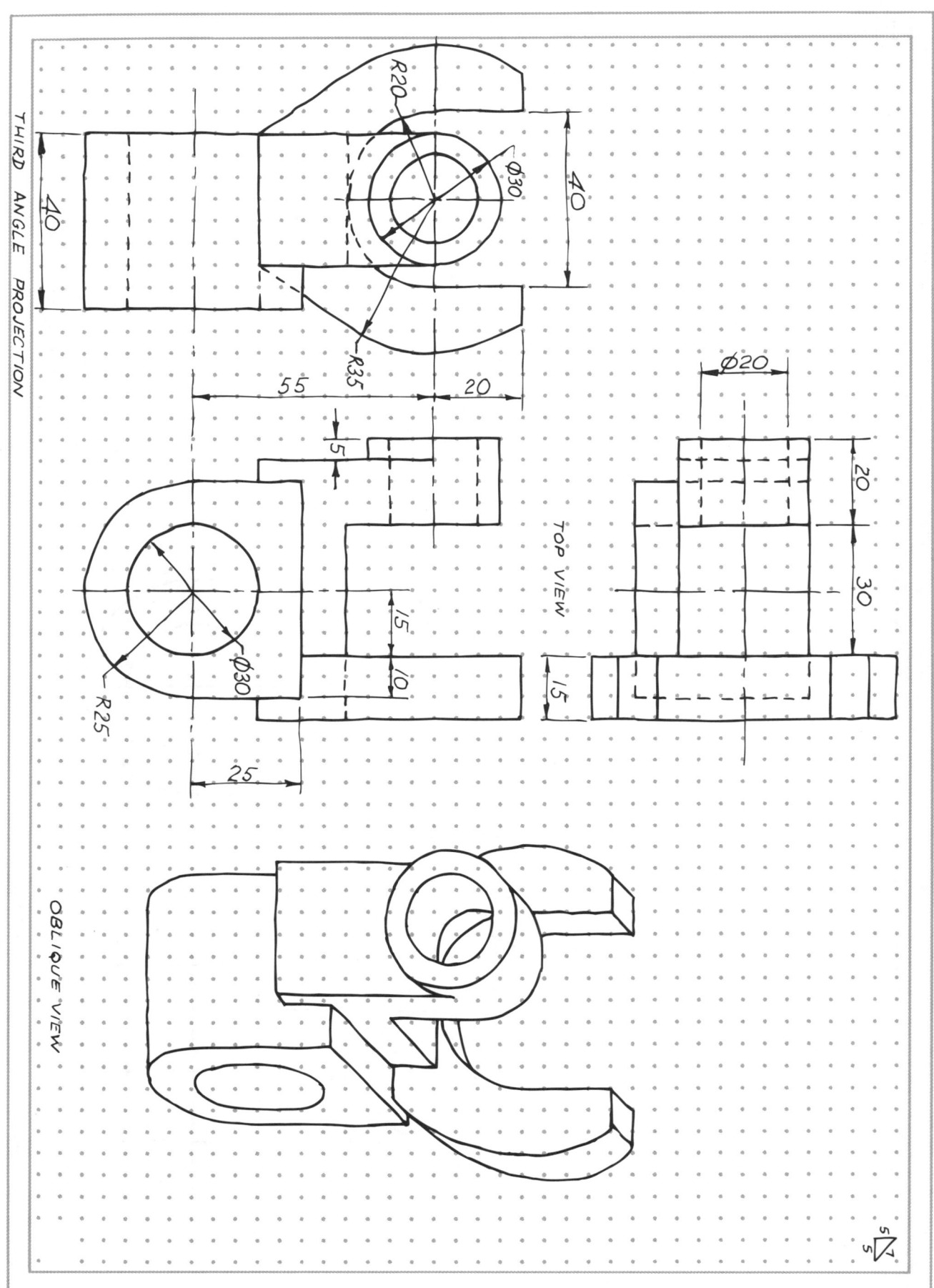

THIRD ANGLE PROJECTION

R20

Ø30

40

40

R35

55

20

5

R25

Ø30

15

10

25

TOP VIEW

Ø20

20

30

15

OBLIQUE VIEW